30647
134

DESERT SAND TO JUNGLE MUD:

Australian Military Vehicles of the Second World War

VOLUME 1:

Trucks, ¼ ton ('jeep') and Trailers

MICHAEL K CECIL AND KEITH WEBB

Published by Image Control Pty Ltd. Melbourne, Australia

Citation: Cecil, Michael K. and Webb, Keith (2023): Trucks, ¼ ton (jeep) and Trailers. Volume 1 of the series Desert Sand to Jungle Mud: Australian Military Vehicles of the Second World War. ISBN 978-0-6458704-0-4

Terminology:

Jeep® is a registered trademark of Chrysler Group LLC.

The term 'jeep' in this publication is used in the historical context of (1) the small ¼ ton four wheel drive truck manufactured by Willys Overland Corporation, the Ford Motor Company, and the Bantam Car Company during the period 1940 to 1945; (2) the usage of the term 'jeep' in many official government documents during the period 1940 to 1945.

Measurements

Measurements are presented in metric unless considered more appropriate to leave in either Imperial or US measure.

The Series Title

The title of this series is 'Desert Sand to Jungle Mud' because it encapsulates the extremes of the environments and terrain where Australian military vehicles were employed during the Second World War. Operational deployments overseas were initially to the desert sands of the Middle East and North Africa. From early 1941, Australian forces began deploying to the arc of tropical islands to the north of Australia, and to the jungles of Malaya, in a vain attempt at discouraging Japan from striking south. By the close of 1941, with the Japanese war machine quickly spreading across the Pacific and South East Asia, the focus of Australia's war effort re-oriented to fighting in the muddy jungles of the region.

Acknowledgements

The authors gratefully acknowledge the assistance of the following individuals and organisations:
Andrew Browne; Ian Austin; Ian Bryce; Ray Everett; Matt Denning; Euan McDonald; Paul Hanson; Lloyd Knight; Alf Brooks; Andrew Carlile; Krystii Melaine; State Library of Victoria; State Library of New South Wales; State Library of Queensland; Australian War Memorial; National Archives of Australia.

ABBREVIATIONS

Abbreviation	**Meaning**
ACF	Australian Comforts Fund
AFV	Armoured Fighting Vehicle
AH	Ampere Hour
AMF	Australian Military Forces
ASC	Australian Service Colour
Aust	Australia or Australian
AWC	Allied Works Council
BCFK	British Commonwealth Forces Korea
BCOF	British Commonwealth Occupation Force
BOD	Base Ordnance Depot
C	Celsius
CDC	Commonwealth Disposals Commission
CMP	Canadian Military Pattern
cwt	Hundredweight
DAFVP	Directorate of Armoured Fighting Vehicles Production
DAP	Department of Aircraft Production
DC	Direct Current
FMC	Ford Motor Company
GMH	General Motors Holden
GOC	General Officer Commanding
H	Height
HE	High Explosive
HP	High Power
KVA	Kilovolt amperes
L	Length
LAD	Light Aid Detachment
lb	Pound
LHQ	Land Headquarters
LP	Low Power
LSM	Landing Ship Medium
LST	Landing Ship Tank
MB-BRT	Model MB British Contract
MGO	Master General of Ordnance
ML	Muzzle Loading
MMG	Medium machine gun
MS	Mild Steel

CHAPTER 1

First Of The Many

CHAPTER 1

Introduction

The 'Trucks, ¼ ton (USA)' proved to be one of the most versatile and useful vehicles used by Australian forces during the war. Authorities in Australia were first alerted to the existence of this new type of vehicle in March 1941, but rejected the vehicle as being an unsuitable type.

The Australian Army later requested an example and the Australian Trade Commissioner obliged by placing American Order 76 directly with the Ford Motor Company in Dearborn Michigan, for a single vehicle at a cash price of $910. In parallel with the Army's request, the Directorate of Armoured Fighting Vehicles Production of the Department of Munitions also requested a single example. This time the Trade Commissioner placed a cash order directly with Willys-Overland for a single vehicle at a cost of $900. Both the Ford and Willys examples were dispatched from the United States early in 1942. They were Australia's first two jeeps. The first quantity delivery of jeeps to Australia under the lend-lease scheme was 500 jeeps, with 250 delivered into Melbourne at the end of June before the ship continued to Sydney to offload the other 250 jeeps in early July. This was the first large consignment of over 15,000 delivered between 1942 and the end of 1945.

Despite the steady arrival of new and reconditioned vehicles from North America, there never seemed to be enough to fulfil all the Army's requirements. While the numbers supplied had risen to 70% of the Army's needs by November 1943, a revision to the Order of Battle in December called for a marked increase in the issue of jeeps and trailers to replace a range of larger vehicles in forward areas. These changes resulted in a much more severe shortage of jeeps with the stocks on hand slipping to only 52% of the Army's revised requirements. Nevertheless, the numbers of jeeps available steadily increased and soon allocations were exceeding war wastage. The number of jeeps and jeep trailers available within the Army at particular dates is given in Appendix 5.

Once a jeep arrived at an Australian port, it was sent to one of several civilian assembly contractors, located mainly in Brisbane, Sydney and Melbourne. Assembly consisted of unpacking, checking, general assembly, tropic-proofing of electrical equipment, repainting overall with the Australian Army basic camouflage colour Khaki Green Number 3 and final assembly including marking the vehicle nomenclature on the underside of the bonnet, engine tuning and road testing. The vehicle was also equipped with the ancillary items supplied by the manufacturer, ready for the final acceptance inspection. Upon passing inspection, the vehicle was allocated a registration number which was then painted on each side of the bonnet and across the rear. Once delivered to the allocated Army vehicle receiving depot, the Australian Army Ordnance-supplied ancillary items were added to complete the Vehicle Outfit List (VOL). The vehicle was then ready for issue to a user unit.

The supply of jeep spare parts to Australia under the Lend-Lease scheme was governed by the number of jeeps supplied and the spare parts annual usage rates calculated by the US Military. These were evidently not modified or increased to reflect the much higher usage rates being incurred in the harsh tropical environments that the Australian Army was operating in across the South West Pacific Area. The extremely poor quality of the tracks, deep mud, numerous water crossings and constant overloading caused a marked increase in the usage of items such as shock absorbers, oil and grease seals, brake components, and springs. To relieve the inconsistent availability of spares, a range of repair parts, including spring sets, windscreens glass, pistons, rings, oil seals and gasket sets were manufactured in Australia.

The 'Trucks, ¼ ton (USA)' remained in service in the post-war period, albeit in much reduced numbers. They were wasted out of RAN service first, with the majority replaced by more conventional vehicles by the early 1950s. The RAAF and the Army maintained a fleet of Ford GPW and Willys MB jeeps until the end of the 1950s and early 1960s, when they were replaced by more modern tactical trucks. In the Army's case, this was the Land Rover, while the RAAF acquired vehicles such as the CJ-5 and International Scout.

A recently assembled Willys Model MB, chassis number MB130878, parked in a Melbourne street.
Registered 79633, the jeep was assigned to MEE at Monegeetta, Victoria, for use in trials and as a general purpose vehicle.
It was used in the towing tests of the prototype 25-pdr Short field gun in late 1942. The jeep was disposed of to Stokoe Motors shortly after the war.

Views of the engine compartment of Willys MB registered 79633, with the headlamps swivelled up and back to provide emergency illumination of the engine.
Note the absence of the blackout driving lamp and guard, indicating the jeep was assembled prior to August 1942.

A member of the test crew stands next to a jeep parked in a swollen creek bed in Central Australia during the hot-dry trials of the Australian Cruiser tank in early 1943. The jeep is marked 'DAFVP' and does not carry an Army registration number on the bonnet. The jeep is likely to be the Willys MB purchased direct from the manufacturer by DAFVP in early 1942, making it the first Willys MB to arrive in Australia.

Some of the early jeep arrivals were disruptively camouflaged in accordance with the mechanical transport orders issued by Army headquarters.
This Willys MB, registered 79549 with chassis number MB130552, appears to be in a three-tone scheme.

The application of disruptive camouflage to ¼ ton trucks was unusual in the AMF, despite its widespread use on larger vehicles. This Willys Model MB, registration 79657, chassis number MB124793, appears to have a disruptive pattern of ASC 'W' Light Earth over the vehicle basic colour of ASC 'J' Khaki Green. The vehicle was serving with the School of Instruction, 2nd Australian Army, at Watsonia Barracks, Victoria, in late 1942. The headquarters of 2nd Australian Army was in the nearby suburb of Ivanhoe at the time.

Shortly after arrival in Australia, Willys Model MB with chassis number MB129786 and engine number MB131325 is seen serving with C Squadron, 2/8th Armoured Regiment, 2nd Armoured Brigade, 1st Australian Armoured Division, in the Narrabri - Wee Waa area of New South Wales. The Division conducted large-scale exercises in the area between August and October 1942. Insects, particularly grasshoppers and locusts, are a common hazard in the area during spring, clogging unprotected radiators and preventing adequate cooling. To prevent this, the unit has equipped the vehicle with a timber frame covered with a sheet of fine wire mesh screen, known as flywire, to intercept the insects before they get to the radiator. The anti-insect screen appears to be bolted to the front panel. The broken remains of an axe, still securely held by the axe head clamp, is of note. This vehicle was possibly deployed to New Guinea with the Regiment in April 1943. It was disposed of at the end of the war by being abandoned by the CDC. This usually meant the vehicle was in a poor state of repair and in a location where sale or recovery was not economically feasible. Note the multiple C Squadron circular markings on the lower windscreen frame, side panel and on the plate attached to the front bumper. A bridge sign yellow disc has been painted onto the far right of the lower windscreen frame but without the numeral '2'.

Assembled, checked and accepted for service, a mix of Willys and Ford jeeps await issue to units from an ordnance vehicle park. The variations in the size of the registration numbers are of note.

Reconditioned ¼ ton trucks engines and transmissions await their final fittings before being packaged for dispatch to an ordnance depot. Keeping up with the demands from field workshops for replacement major unit assemblies was made more difficult by the sometimes restricted allocation and supply of spare parts from the United States. To help relieve the situation, many items such as pistons, rings and gasket sets were manufactured in Australia. Reconditioning of major assemblies was initially carried out by Army workshops, but to help service the high demand and overcome the backlog, civilian automotive companies were also contracted to recondition engines, transmissions and axle assemblies.

Craftsman W.T. Hartley completes an oil filter change on a jeep at an ordnance vehicle park in Tolga, Qld in January 1944.

The jeep engine is the configuration equipped with a PCV valve on the manifold re-breather connection between the oil tube and the air tube to the carburettor. In August 1942, the Australian Army introduced the requirement to paint an abbreviated form of vehicle nomenclature within the engine compartment. On jeeps, this was invariably on the underside of the bonnet, as seen here stencilled in white letters and numerals. Given that there was no practical difference between the two suppliers, Willys and Ford, the basic nomenclature rarely included the manufacturer.

There were many interpretations of the under-bonnet nomenclature requirement leading to variations in the form and information included. For example, the inclusion of either 'MB' or 'GPW' either before or after '¼ ton' as seen here on a preserved example. The stencilling shows some interesting differences including 'GPW' in a larger sized letters and an offset between the two '4' numerals in '1944' as if the last '4' was stencilled separately to '194'.

Included in the Australian VOL for jeeps was the 'Instruction book Driving and Maintenance' first published in July 1944. It was 190.5-mm x 127-mm x 6.35-mm and provided the basic information necessary for the correct and economical operation of the vehicle. The information within the book was extracted from the various manufacturers' manuals but converted to Imperial measure and modified for Australian standards.

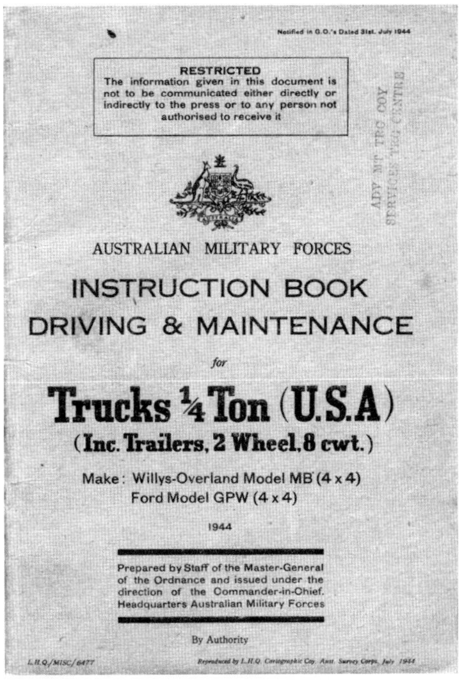

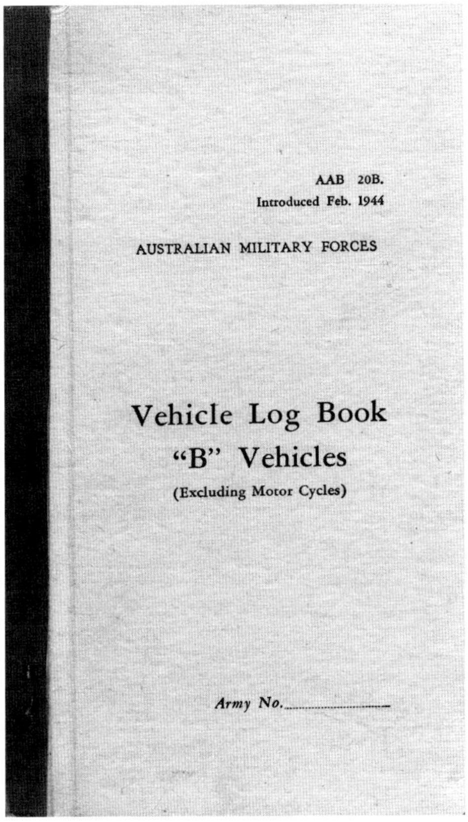

The VOL also included a log book and a steel log book holder. The log book had provision to record the daily operation and maintenance of the vehicle, including distance travelled, fuel and oil used, and any maintenance or repairs carried out. Although the steel log book holder was supplied with the VOL and designed to be mounted in a convenient location on the vehicle, it appears that most log books lived inside the glove compartment or under the passenger seat.

In September 1944, the Australian Army introduced a 'Plates, Vehicle Identification' for all non-armoured vehicles. It was an etched brass plate measuring 76-mm x 54-mm x 0.95-mm. The plate listed the nomenclature, manufacturer's model number, the MVFS number and the Army Census number, the data being either hand stamped or etched. All jeeps had the nomenclature 'Trucks ¼ ton USA' and the Army Census number 6010. The MVFS number was unique to each manufacturer, with Ford being 1027 and Willys being 1028. This plate incorrectly shows the Willys Model MB with the Ford MVFS number. The plate was normally attached to the dash panel in a convenient location, usually to the right of the gauges, using plain round head drive rivets.

CHAPTER 2

In Service

The jeep soon became an indispensable part of the Australian Army, replacing motorcycles and a range of light vehicles in front line units. By February 1945, jeeps dominated the transport on issue within 1 Aust Corps, with 1,785 jeeps and 1,480 8-cwt jeep trailers on issue. They had entirely displaced the 15-cwt 4x4 truck in front line service, with only 33 still in use within the 1 Aust Corps Base Sub-Area. It was little wonder that the ¼ ton truck was so highly sought after. These 'go-anywhere' vehicles were simple, rugged, reliable and minor breakdowns were easy to repair. Their small size and manoeuvrability made them easier for a group of soldiers to extricate when bogged, and belied their ability to carry or tow heavy loads.

Given the extraordinary performance of the jeep, it is little wonder that field units were so reluctant to relinquish their hold on damaged jeeps as there was no guarantee that an immediate replacement would be forthcoming. As a consequence, field workshops and Light Aid Detachments went to extraordinary measures to repair jeeps that would normally have been sent back through the logistics system for repairs by higher level workshops or written off.

In addition to the 'Trucks, ¼ ton, (USA)', Australia also received a small number of the 'Trucks, ¼ ton Amphibious (USA)' – the Ford Model GPA. These arrived in several shipments between June and September 1943, and were discharged in both Melbourne and Sydney. By October, 161 had been delivered and were at the assembly contractor's works. However, adverse reports had been received from US authorities about defects in the GPA, including a weak hull that was susceptible to leaks, the small ground clearance, excessive weight and a lack of power in land operations. Local testing in Australia had also revealed the GPA's poor performance in surf and fast-moving streams – both environments in which the GPA was expected to operate in support of amphibious landings and when crossing creeks prior to the construction of bridging. In the opinion of one Australian soldier, the GPA was only good for taking the generals ashore and the nurses fishing – and then only in calm conditions. As a consequence, further shipments were cancelled and it was recommended that the GPA be restricted to land operations within Australia. In July 1946, it was determined that there was no post-war requirement for the GPA, and all vehicles were sent for disposal.

A happy bunch of Canadians hitching a ride to the Mess for lunch at Watsonia Camp, Victoria. The Ford Model GPW carries the LHQ formation sign of the AMF's rising sun badge over a boomerang in white on a black background. The Canadians were radar specialists stationed at Watsonia as part of the Army's signals establishment.

An early production 'slat grille' Willys Overland Corporation Model MB-BRT, chassis number MB101448, being driven by Major Mitchell at Fossil Downs Station during an extensive survey of the remote Kimberley Region of northwest Western Australia. Mitchell had driven the jeep across the dry, sandy bed of the Margaret River, some 50 kilometres to the southwest near Fitzroy Crossing, to reach the Fossil Downs homestead. Registration number 129219 was assembled and taken on charge in NSW.

Movie director Charles Chauvel uses the sand to map out what he wants all the extras played by serving military personnel to do in the scenes being filmed in the sand hills at Cronulla, NSW, on 7 September 1943. The film 'The Rats of Tobruk' depicted the defence of Tobruk in North Africa during 1941. The filming was extensively supported by the Australian services. While the extras are in Khaki uniforms with steel helmets and 1937 pattern web equipment, the liaison officer is dressed in jungle green. He is leaning on the bonnet of his Willys MB, registered 129628 with chassis number MB227165.

The jeep sacrificed driver comfort for a compact size. The Australian Army recruit during the Second World War ranged between 1.7 and 1.74-metres tall, with the average being 1.72-metres. Most jeep drivers had to crouch a little to see below the upper windscreen frame, and the short distance between the seat and pedals resulted in his legs always being bent upwards. Once the windscreen frame had been bent backwards a few degrees, the canvas canopy could no longer be held tightly, which resulted in it 'flapping' against the top of the driver's and front passenger's heads. Despite the shortcomings, the jeep was an indispensable component of Australia's military transport during and immediately after the Second World War.

A 'Trucks, ¼ ton (USA)' in its element – negotiating rough and rutted cross country tracks. This was part of the rough cross-country course regularly used for vehicle trials by MEE located at Monegeetta, in Victoria. The course, much of it through nearby publicly-owned State forest, included grades up to 16%. A number of short-radius climbing corners and areas of loose, shattered granite tested all types of vehicles to the limit. This Ford Model GPW, registration 133186, chassis number GPW128337, served at MEE for much of the war. It remained in Army service post-war, receiving two engine changes during its latter period of service before being written off at 4th BOD on 13 July 1956. Note the absence of the front bumper bar and the rear towing pintle.

On the beach in the Coorong, South Australia, is an odd assortment of people with their Ford GPW, registration 132511. The driver and the chap sitting awkwardly at the rear are Army personnel, while the man sitting on the bonnet is RAN. They are accompanied by a civilian wearing a knitted vest. The jeep is fitted with chains to negotiate the soft dune sands on the way to the beach and appears to be carrying the crew's sleeping swags. The jeep, chassis number GPW127207, was taken on charge in Victoria, but ended its military career in South Australia in mid-1956 when it was disposed of from 4BOD.

Personnel of the Services Reconnaissance Department (SRD), the Army's unit that specialised in actions behind enemy lines, practising on a parachutists' exit ramp simulator. A similar ramp was installed in the fuselage of a B-24 bomber for parachuting personnel and equipment over enemy territory. Their jeep, a Ford GPW, either registration 130535, chassis number GPW143352 or 160535, chassis number GPW204044, is unusual as it is fitted with white wall tyres – an embellishment not normally seen on Australian jeeps during the Second World War.

Students trying to extricate their stranded jeeps from an old creek line cut through the bed of the Hume Reservoir. The Victorian Lines of Communication MT Training Depot instructors took advantage of the unusually dry bed of the reservoir in May 1945 to conduct part of an NCOs driving and servicing course. Note the round yellow bridge disc and black numeral '2' on the lower windscreen frame, and the unit sign plate of white numerals on a black background attached to the front bumper.

A jeep at large in Central Australia. This is the Willys MB purchased by DAFVP which was accompanying the Australian Cruiser tank Mk.1 during the hot-dry trials conducted in early 1943.

Student signallers from 5th Australian Field Regiment practise using WS No.22 (Aust) receiver/transmitters under the watchful eyes of their instructors. Parked in the background is a jeep with a US military pattern 8-cwt trailer. Note the absence of the blackout driving lamp and guard from the front left mudguard, and the jerry can holder from the rear panel, indicating an early production vehicle.

A Ford GPW at Oro Bay, New Guinea, showing signs of extensive use. The front bumper and edges of the mudguards are battered and bent, and the canopy bow and attaching brackets are missing.

A race against time and tide! Australian and US troops work to extricate the stranded vehicles. With a traffic jam further inland, these vehicles were unable to clear the beach before being caught by the incoming tide in the Lae-Salamaua area of Papua New Guinea. Both the jeeps are equipped with wireless. The near jeep is a Ford GPW, registration 128023, chassis number GPW78966, serving with the divisional artillery headquarters. Note the aerial base attached to the rear right corner of the body and the folding tables lashed to the roof and rear of the canopy. The jeep further forward appears to be part of a heavy wireless unit operating the WS No.133 (Aust). Note the aerial pole sections mounted in brackets on the left side of the body.

A jeep serving with a Light Aid Detachment (LAD) crawls across a track corduroyed with logs cut from the surrounding forest. Tracks soon became quagmires with the repeated passage of vehicles, so covering with logs was a quick solution, but the result needed to be negotiated slowly to avoid damage to the vehicle.

A jeep named 'Skeeter' approaches a guard post at a cross road in Dutch New Guinea. The bridge sign on the right lower windscreen frame is an irregular shape in accordance with the official markings instructions. The unit sign on the left of the lower windscreen frame has been obscured by the censor. The guard is armed with a No.1 Mk.3* rifle with a fixed Pattern 1907 bayonet.

Seven soldiers crammed into a four-seat jeep driving through a village 'somewhere in Papua New Guinea'. Note the windscreen has the 'Covers, Canvas, Windscreen' in position to prevent light reflection. Although included in the VOL issued in July 1944, it was deleted from the VOL issued in November 1945. (See Appendix 1)

Logs cut from the surrounding jungle were often used to corduroy particularly bad sections of muddy tracks. The driver of this Ford GPW, registration 147131, chassis number GPW178915, watches carefully as the front wheels start over the logs. Other jeeps wait their turn further back. Note the jeep has had the entire windscreen frame removed, and the windscreen hold down catches from the front of the bonnet relocated to the back to prevent the lower back edges of the bonnet from vibrating. This jeep remained in service post-war, being written off by Board of Survey at Detachment 3rd BOD, Seymour, Victoria in July 1956.

Gunners of 2/6th Australian Field Regiment aboard a 'go-anywhere' jeep equipped with wheel chains, drive through a sea of liquid mud in Borneo in 1945.

Jungle mud that even a ¼ ton truck has trouble negotiating! The crews of these two vehicles try and extricate a jeep from a deeply rutted, muddy jungle track, while a short-wheelbase 2 ½ ton GMC GS truck awaits clearance of the 'road'. In many cases, jeeps were the only vehicles able to maintain the flow of supplies forward and to carry wounded back from the front line along narrow, boggy jungle tracks. They proved to be the most versatile and capable vehicle used in the jungle warfare conducted by Australian forces in the latter period of the war. Consequently, units were reluctant to relinquish their jeeps no matter how battered they became.

A trio of jeeps moving through a village in Papua New Guinea in 1944. The leading jeep has had the registration number blanked out by the censor. Note the canvas cover over the folded windscreen to prevent glare and light reflection.

Overloaded with Australian and Dutch troops, a jeep drives under an ornamental archway in a recently-liberated area of Dutch New Guinea. The crest at the top of the gate has the words 'Je Maintiendrai' which translates to 'I will maintain', while the main slogan on the arch is 'Nederland Zal Herrijzen' – Netherlands will rise again. The jeep carries no markings and has had the windscreen frame removed.

A jeep with wheel chains fitted drives across a creek near Finschhafen, Papua New Guinea while engineers construct a substantial log bridge. Come the wet season, creeks such as this became impassable, making the delivery of supplies to forward areas impossible without such crossing improvements.

The more traffic, the more the jungle tracks became a sea of mud. In the Finschhafen area of Papua New Guinea, the driver of Willys MB registered 127756, chassis number MB197453, stops for a chat with a fellow soldier. Curiously, the jeep is fitted with a tyre chain only on the left side front wheel, rendering it useless once the right side wheel loses traction. Note the pattern of the inner springs on the surface of the seat cushions and that there is no blackout driving light fitted to the left mudguard.

The mud of Finschhafen! The tracks along the coastal strip of Papua New Guinea were soon turned into a quagmire by the vehicle traffic. This jeep, registered 128023, is a Ford GPW, chassis number GPW78966. It is fitted with a canvas cover over the windscreen to prevent light reflection. It has chains fitted to all four wheels and a steel wire tow rope wrapped around the front bumper and chassis rails. Another jeep can be seen in the background heading in the opposite direction.

Taking the opportunity to get the mud off their jeep in a creek on the Huon Peninsula, Papua New Guinea. A tactical marking of 'B2' is partially visible on the side. The spare wheel assembly is either unbolted from the mount to assist with cleaning, or has been badly bent in an accident.

A gaggle of vehicles, both Australian and American, take advantage of a coastal creek to get the accumulated mud off. The poor state of the tracks and the constantly muddy conditions encountered in tropical service resulted in a much higher consumption rate of fast wearing spares such as axle seals, universal joints, shock absorbers, suspension springs and brake linings than originally estimated by the logistics services. When supplies from the USA could not keep up, the Australian Army turned to local manufacturers to make up the shortfall.

The driver of this Ford GPW registered 130647, chassis number GPW130451, carefully negotiates a rudimentary palm log bridge. The jeep appears to be recently issued as there are none of the fair-wear-and-tear dents and scratches that quickly accumulated when operating in close jungle terrain. The unit sign of 134 on a red over blue background with a 1-inch white band across the bottom indicates the jeep is operated by a field artillery unit attached to a Corps, but without a visible formation sign, it is not possible to ascertain Corps and unit.

A well-laden jeep and trailer negotiate a temporary pontoon bridge over a coastal creek in Papua New Guinea. The trailer is a US manufactured type (Willys or Bantam) that has had the mudguards removed. The jeep is fitted with a set of wheel chains. It is missing the windscreen and canopy bows and has a 2-Imperial gallon fuel can in place of the more usual 5-US gallon jerrycan.

Engineers haul on ropes to pull a ferry made from Arc-mesh and pontoon boats across the Gilman River in New Guinea. The Ford GPW registered 155135 with chassis number GPW165220, is towing a US manufactured trailer. The vehicles belong to the 101st Brigade Support Company which was engaged in experiments in laying smoke screens using 4.2-inch Heavy ML SB mortars. Note that the entire windscreen frame has been removed as well as the canopy bows. The jeep remained in service until late 1959.

Staff of 8th Aust Movement Control Group (Type D) standing around their Ford GPW, registered 134745, chassis number GPW165336, at the airfield at Finschhafen, Papua New Guinea in late September 1944. The comparatively small size of the registration number and the large 'AIF MOVEMENT CONTROL' are of note. The jeep is showing signs of considerable use, with the standard front bumper replaced by a piece of steel tubing, while the front edge of the right mudguard and the grill are battered and bent. The canopy pulled down over the top edge of the windscreen is unusual.

A lieutenant smokes a cigarette while he and his driver contemplate a road sign in the Finschhafen area of Papua New Guinea. The jeep windscreen is covered with what appears to be the jeep's canvas canopy to prevent light reflection.

This very neat and well-cared for Ford GPW, which has the registration number of 130339 and chassis number GPW114932, being used to take a group of unit personnel to the beach. The unit using the stylised insignia on the side has not been identified.

There was not much padding in a jeep seat base!

Two jeeps on a raised ramp for servicing. Both jeeps have the War Office unit serial number 45732 in white over the three colour bars of yellow over blue over yellow, painted on the side panel just behind the front mudguard. Together with the unit sign of 2 over 107, in white on a black background, the signage indicates these jeeps are serving with 2 Aust Corps HQ. The large 'CAMP' sign possibly indicates the vehicle is for local area use only. The under-bonnet nomenclature on the far vehicle has 'QT-WILLYS/43', an unusual interpretation of the required nomenclature. The near jeep is registered 136735, and is a Willys MB, chassis number MB231605. It was transferred to the PMG Department in the immediate post-war period.

A Willys Model MB-BRT, registration 128793, chassis number MB227879, serving with a workshop unit in Northern Australia. In the background, several CMP 'Trucks, 3 ton, Machinery' can be seen set up with their canvas covers extended as awnings to provide the maximum covered workspace. Towards the end of the war, the jeep had displaced several classes of vehicles in front line service, including cars, motorcycles, vans and 15-cwt trucks. This is one of a couple of thousand 'Trucks, ¼ ton (USA)' that remained in service with the post-war army. It was written off by Board of Survey at Detachment 2nd BOD in September 1960.

VX86142 Lance Corporal Matthew 'Matt' Bickham with his ¼ ton truck 'Fluff' in July1945, while serving with 1st Australian Mobile Lithographic Section. The vehicle, a Willys Model MB with registration 160116, must have been well looked after by the unit, as it survived the disposals of the immediate post-war period. The engine was replaced in 1948, and 'Fluff' continued in service until September 1960, when it was written off at 2nd BOD in Sydney. Personal names on vehicles were reasonably common, though not always as neatly sign-written as 'Fluff'. The Formation sign of a white rising sun over a boomerang on a black background painted on the right side of the lower windscreen frame is that of Army Headquarters. On the left side of the lower windscreen frame is the unit sign of a blue background superimposed with '761' in white.

The left side of 'Fluff' showing the War Office Unit Serial number '45201' and three colour bars on a white rectangular background. Unusually, this marking was painted on both sides.

The right side of the jeep body has three sets of markings. Closest to the rear wheel arch is the tare weight marking 'T C Q', standing for tons, hundredweight and quarters, with the figures '1 2 0' below, so a total weight of 2,464-lbs Imperial. The official weight for logistics purposes was 2,352-lbs Imperial. Roughly painted below that is 'TARE 1 TON' leaving no room for misinterpretation! Forward of that is, from top to bottom, the height (H), length (L) and width (W) in feet and inches. This is the shipping height with the canopy stowed, windscreen folded forward and the spare wheel dismounted and stowed flat inside the body. The official dimensions for logistics purposes were 4-feet 3-inches high, 5-feet 6-inches wide and 11-feet 6-inches long. Further forward and obscured by Matt's leg is the War Office Unit Serial painted onto a rectangular white panel. The unit serial number '45201' appears to be painted in a dark colour with the three colour bars of, from top to bottom, dark brown, red, dark brown.

Note the mixture of Khaki and Jungle Green uniform items, and the different gaiters – the officer at left wearing the earlier web anklets, while the one standing at right has the later canvas 'jungle' gaiters.

A local of the Wewak, Papua New Guinea area stands on the roadside adjacent to a jeep operated by the 8th Docks Operating Company (Type C) AIF, in June 1945. The unit sign is an 8 over 733 on a blue background, while the formation sign of a rising sun over a boomerang, in white on a black background indicates the unit is attached to Land Headquarters. The rear canopy of the jeep has been rolled up which, when combined with the open windscreen, provided good ventilation while still protecting the occupants from the sun and rain. Behind the jeep is a 3-ton CMP truck operated by the 1st Field Company, RAE, indicated by the unit sign 1 over 46 on a blue background.

A line of jeeps and trailers of various 9th Australian Infantry Division units awaiting orders to move aboard their allocated LST during preparation for the invasion of Borneo in 1945. All the jeeps have been waterproofed for deep wading with vertical extensions fitted to the exhaust pipe and carburettor. To fit the carburettor extension tube, the bonnets have been removed and are stowed at the rear between the spare wheel and canopy. The 8-cwt trailers are mostly of US manufacture, with two Australian No.3 trailers near the front of the line, one with a modified half-round raised canopy frame.

Lashing down a jeep and trailer on an Australian small craft in preparation for movement. Several Army small craft are docked together while preparations for sea are completed. Each small craft is armed with a 'Small Craft, .303-inch, Twin Vickers MMG' mounting for anti-aircraft defence.

Landing vehicles from larger transport ships often involved slinging into a landing craft using the ship's crane, in this case, during landings on the island of New Britain. The jeep trailer has already been loaded aboard and is being followed by the jeep. Note that the jeep is not fitted with a blackout driving lamp and guard on the front left mudguard. There is also a non-standard fuel can carrier on the rear panel manufactured from steel strap. The lower windscreen frame has a fire extinguisher bracket mounted horizontally on the right and a first aid box on the left.

Looking forward along the upper deck of an LST crowded with 9th Australian Infantry Division jeeps, trailers, men and piles of personal kit. There are multiple 40-mm light anti aircraft guns, all loaded with a clip of ammunition and all pointed outwards for immediate air defence.

Disembarking from an LST onto the north coast of Papua New Guinea in the Lae-Salamaua area. The jeeps are fitted with wheel chains and are being loaded with additional boxes of equipment from within the LST. At the left is a 2-pdr Tank Attack gun. Although the little 2-pdr was soon outclassed in the North Africa and European theatres, it remained a very effective weapon in the SWPA, particularly once the Australians developed a nose-fuzed HE round and a Canister anti-personnel round. The towing vehicle late in the war was often a jeep. In the foreground is a trailer loaded with an excavator bucket. The censor has erased the identifying numbers from both the LST and the jeep.

Engineers built or improved the roads and tracks in Papua New Guinea as Allied forces moved steadily forward. This road through the Finnesterre Ranges was built to allow larger vehicles to bring supplies forward and casualties back. The engineers in this jeep have stopped while waiting for the grading of the track at the corner. Note the steel canister attached to the back of the driver's seat with a quick release strap and catch.

Two Diggers ponder the complexities under the bonnet of their 'Truck, ¼ ton' beside a well-constructed native hut and food storage. The vehicle has been modified for fitting the deep fording kit, as evidenced by the round hole visible in the lower part of the bonnet which fitted the carburettor snorkel extension. Wrapped around the front bumper bar is the 'Ropes, Towing, Steel, Medium (Aust) No.1' which was a standard item in the Australian Army's VOL for the ¼ ton truck. The 'Handles, Starting' is also visible protruding from the access aperture in front bumper – further evidence that the vehicle is suffering from a mechanical problem. It is towing a US-type 8-cwt trailer, characterised by the rounded mudguards.

The Australian Comforts Fund (ACF) was a civilian umbrella organisation that channelled patriotic donations in the form of comforts that is, items or services in excess of those normally supplied, to serving personnel. Various non-government organisations (NGOs) such as the Young Men's Christian Association (YMCA), the Young Women's Christian Association (YWCA) and the Salvation Army worked under the auspices of the ACF. The comforts provided ranged from cups of tea and a biscuit through to rest stops and film showings. The NGOs also had Field Commissioners embedded with units. The commissioners were given an honorary officer rank and were supported by the military logistics system, including the provision of appropriate transport, such as this jeep assigned to a YMCA field commissioner.

The ACF field commissioners were always a welcome sight to front-line troops. Soldiers on Labuan Island crowd around the YMCA field commissioner's jeep to have their mugs filled with a hot cup of tea. The commissioner was embedded with 2/28th Australian Infantry Battalion, as indicated by the War Office unit serial number 61189, just visible on the left side of the scuttle. The three colour bars beneath the sign are khaki green over white over khaki green. The Latin phrase 'ICH DIEN' translates as 'I Serve'. Faintly chalked on the lower windscreen frame are the embarkation control numbers indicating which vessel and loading position the jeep was to travel aboard.

A USAAC C-47 thunders into the air from the main airstrip at Port Moresby, New Guinea. In the foreground are jeeps and trucks waiting to cross the airstrip to the camp area on the far side. The nearest jeep is a Willyss MB-BRT, registration 76951, chassis number MB136534.

Private B. McGrory contemplates a road sign in New Guinea while Private G. Troop looks on. The large 'Ferry Service' sign is secured to the top edge of the windscreen frame by two flat sections of strap that slip into two brackets attached to the windscreen frame on each side. Hence, it can be lifted off if the jeep is engaged on other duties.
The unit operating the ferry service is unknown as the censor has obliterated all the identifying signage. The jeep appears to have a coloured stripe, possibly yellow, of about 150-mm in width painted across the scuttle.

A very grainy image of what appears to be registration 131276 parked beside a muddy track. The placement of the registration across the chassis member below the pintle hook is unusual. The jeep is fitted with tyre chains for additional traction. Note that the bumperettes have been mounted upside down.

Crossing a river in the Maprik area of Papua New Guinea are jeeps and Australian No.3 trailers of the 6th Australian Infantry Division. The jeeps have had the windscreen frames removed. Note how the No.3 trailer does not sit level behind the jeep. This was recognised as one of the shortcomings of the design.

Despite the tyre chains, registration 127696, a well-laden Willys MB with chassis number MB195894, needs a tow from an engineer bulldozer to climb the bank of the river in the Finschhafen area of Papua New Guinea.

A jeep and trailer getting some assistance down a steep slope to the river from a bulldozer. The jeep trailer has had the mudguards torn off recently as they are visible on top of the load. The dozer carries the registration AWC17018. It was not uncommon for heavy earthmoving equipment to be transferred from the Allied Works Council (AWC) to the Army and Air Force for operations.

Multiple manpower and a couple of gun ropes give this well-laden 5th Australian Infantry Division jeep and Aust No.2 trailer a helping hand up the steep and muddy bank of a river in Papua New Guinea. Note that the jeep has had the entire windscreen frame removed.

Despite the chains on the rear wheels, this jeep slid off a jungle track in the Sattelberg area of Papua New Guinea and flipped over, damaging panels and flattening the steering wheel. Soldiers have righted the vehicle and are winching it back onto the track using pulleys, ropes and logs cut from the jungle. Accidental damage in such conditions was frequent. Note that the censor has obliterated the jeep's registration number.

Even well-graded and used tracks could be dangerous. A soldier armed with an Owen Machine Carbine surveys the result of a Japanese land mine placed on the Madang-Finschhafen Road, Papua New Guinea. The explosion blew a crater on the edge of the road and flipped the Ford GPW, registration 58091, chassis number GPW113077, and trailer into the bush on the far side of the road.

The harsh tropical operating environment accelerated the wear and tear on mechanical transport of all types. With jeeps being an almost indispensable vehicle in forward areas, coupled with the difficulty of getting replacements and even fast wearing spare parts, units and field workshops went to extraordinary lengths to retain and repair jeeps and return them to service as quickly as possible.

Functional improvements were also implemented in an effort to overcome persistent failures of some parts, such as differential seals, the steering bell crank and vehicle springs. The installation of heavier, more robust vehicle springs was officially implemented in January 1945, with front springs increased from eight to ten leaves, and rear springs from nine to eleven leaves.

Sometimes, the damage was so extensive the vehicle was only good for salvage and providing spare parts to keep other jeeps in service.

Two Australian soldiers, one armed with an Owen Machine Carbine, take charge of an emaciated and demoralised Japanese soldier. Their jeep, visible in the background, has had the bumperettes mounted upside down.

Two Australian Military Policemen escort a bare-foot wounded Japanese prisoner to their Ford GPW, registration 128086, chassis number GPW65230, on Borneo in 1945. A second wounded prisoner is visible seated behind the vehicle's driver. The jeep is showing all the usual signs of hard use – bent mudguards and a damaged front bumper. The vehicle is equipped with a manually operated spot-lamp on the driver's side – a common addition to vehicles of Military Police units. This GPW was not returned to Australia at the end of the war, but was transferred to 50th Civil Affairs Unit, Borneo on 6 February 1946.

Elements of the Japanese surrender party unload baggage from a Mitsubishi Ki-21 Sally heavy bomber into the back of a jeep operated by the Provost Company, 9th Australian Infantry Division on Labuan Island in September 1945. The Sally is painted white with large green crosses in accordance with the Allied requirements for aircraft conveying surrender parties in the SWPA. The main surrender party led by Lieutenant General Masao Baba had arrived in a second aircraft, a Tachikawa Ki-54 Hickory transport aircraft.

Lieutenant General Hatazo Adachi, commander of the Japanese 18th Army being driven to the surrender ceremony at Cape Wom, Papua New Guinea, on 13 September 1945. The jeep, registered 58091, is a Ford GPW, chassis number GPW113094. The formation sign on the right side of the lower windscreen frame is the kangaroo over a boomerang, in white on a black background, of the 6th Australian Infantry Division, while the unit sign on the left side, a 6 over 84 in white on a black background, is that of the 6th Australian Infantry Division Headquarters. The surrender was accepted by Major General H.C.H. 'Red Robbie' Robertson, GOC, 6th Australian Infantry Division. Lieutenant General Adachi was required to surrender the sword he can be seen carrying.
(The sword is held in the collection of the Australian War Memorial.)

An unidentified Japanese senior officer waits to be driven to a surrender ceremony. The formation sign painted on the lower windscreen frame of the jeep is that of the Australian Land Headquarters.

TX2244 Private Cecil Doran of 2/8th Transport Platoon bidding Warrant Officer Jim Prowse and Privates Jim Newman and Bill Cauldwell farewell as they start their journey back to Australia and home. The three men in the jeep have qualified for early repatriation under the 'five years of service' rule. As there was no provision to keep the canopy bows taut when the rear of the canopy was rolled up, a piece of twine has been used.

Personnel from Australian Army Ordnance Corps field units take a break on a designated nude swimming beach in Papua New Guinea. Transport to the beach included a Ford GPW, registration 130393, chassis number GPW111990.

A jeep serving with an Army Headquarters engineer unit during the 1950s. The vehicle has been repainted in gloss Deep Bronze Green paint, with the registration number stencilled on either side of the bonnet. Note the damaged lifting handles on the side and rear, and the use of a rectangular convex rear vision mirror sourced from a Canadian Military Pattern vehicle.

Australia's post-war contribution to the British Commonwealth Occupation Force (BCOF) in Japan included the 34th Australian Infantry Brigade Group with its full compliment of equipment. The occupants of this jeep, a Ford GPW registration 155223, are seemingly oblivious to the local boys checking out the contents of the rear seating area. The jeep carries the BCOF formation sign of a King's Crown over the letters BCOF in the centre of the lower windscreen frame. The jeep has several interesting features. The top edge of the windscreen frame had an additional piece added to increase the headroom over the driving compartment by about 100-mm and turnbuckles have been fitted to the windscreen frame and body to secure demountable canvas doors. The windscreen hold-down catch and retainer have been relocated from the front of the bonnet to the rear lower corners to reduce the vibration of the rear edge of the bonnet. The seat pads also appear to be substantially thicker than those originally issued.

A jeep operated by 22nd Field Regiment (Self Propelled) parked outside the drill hall in Landcox Street Brighton, Victoria, in the early 1950s. The jeep remained the Australian Army's primary light tactical truck until the introduction of the short wheelbase ¼ ton Land Rover Series 2 in the late 1950s. This jeep appears to be registered 112785, with the numerals painted across the rear bumperettes, however that registration was first issued to a Land Rover Series 2A in 1963, so the actual registration of the vehicle is unknown. Just visible on the rear right of the canopy is the 'CAUTION LEFT HAND DRIVE' sign – a necessity in a country where vehicles are configured with right hand drive and are driven on the left side of the carriageway.

Lieutenant General H.C.H. 'Red Robbie' Robertson inspecting servicemen at Puckapunyal in the 1950s. The 'Truck ¼ ton (USA)' was retained for service with the post-war Army, with the last ones being disposed of in the early 1960s. The vehicle is finished in gloss Deep Bronze Green paint. The brilliant white seat covers and location of the Pyrene fire extinguisher adjacent to the shovel bracket are interesting points. The fitting of the fire extinguisher in this position precludes the carriage of the shovel. Personnel are dressed in the 1950s wool-serge battledress with webbing anklets. The officer is equipped with a 1937 Pattern web equipment pistol holster and ammunition pouch, while the soldiers all have 'Rifles, .303 inch, No.1 Mk.3*' with fixed 'Pattern 1907' bayonets. Registration 79624 was a Willys Model MB with chassis number MB130614 and original engine number MB133110. It was fitted with a replacement engine, MB230379, during service. It was written off at Detachment, 3rd BOD, Seymour, on 17 December 1959, and was listed for disposal on 5 January 1960.

Three personnel from the Southern Command General Transport Company aboard their Ford GPW, registration number 161902, during the 1950s. The vehicle is finished in the post-war gloss Deep Bronze Green colour and is fitted with tyres similar in tread pattern to the Dunlop 'Silent 90' brand designed for road use. Note the remote turn 'trafficator' indicator assembly bolted to the inside face of the lower windscreen frame – a necessary addition for a left hand drive vehicle being driven in a country where civilian vehicles were right hand drive and driven on the left side of the road. The trafficator arm allowed the driver to indicate his intention to either stop or turn by extending a yellow metal hand-shape from within the rectangular steel cover by pulling a knob on the end of a Bowden cable or steel rod. Note the recruiting poster board mounted on the side. It was common for CMF units to solicit recruits directly into the unit through such advertising. The chassis serial GPW20586460 indicates the vehicle was manufactured during June 1944. It was disposed of from 4th BOD in South Australia during November 1959. Visible in the background on the left is a 'Truck, 2 ½ ton, GS (USA)' – a GMC CCKW-353 with a soft-top open cabin, and beyond that to the right is a 'Buses, 21/26 Seater (Aust) No.2'. The No.2 bus comprised an 'austerity' bus body fitted to a Dodge T110-L5, 4x2, 160-inch wheel base chassis.

The front left view of 161902 showing several interesting features. A large brass hand-held fire extinguisher has been mounted to the exterior in a position which prevents the stowage of a shovel. The bonnet has an additional hold-down clip at the rear on each side, and the windscreen hold-down clips have been removed, as has the blackout driving lamp and bracket from the front left mudguard. The left side rear vision mirror and bracket have been removed, with the mirror remounted to the upper centre of the windscreen. The windscreen has a manual wiper on the right side and a vacuum-operated wiper from a CMP truck on the left. Note the high-gloss Deep Bronze Green finish evident on the wheels. The jeep was repainted with the windscreen in the upright position, as the original lustreless colour can be seen where the windscreen sealing rubber masked the bodywork. In the background are several CMP 3-ton GS trucks, another jeep and a 'Truck, 6-seater Utility' based on a Ford F15 chassis. The front of a 'Truck, 3-ton, Ambulance (Aust) No.2' can be seen at the left of frame. Although built on a 30-cwt capacity F30 chassis, the Australian Army classed the ambulance as a 3-ton vehicle with 'special springing'.

Front right view of 161902 less the passengers. The outline of the windscreen frame in the earlier Khaki Green No.3 paint is clearly visible from when the jeep was repainted in Deep Bronze Green with the windscreen in the upright position.

Recruitment advertising for the Australian Regular Army (ARA) was also carried on jeeps. Registration number 128276 is a Willys Model MB, as is 7939?, which could be any of the ten jeeps in the registration range 79390 to 79399. Registrations 162483 and 159669 are both Ford Model GPW. The jeeps with the bonnets visible have all had the windscreen frame hold-down clips removed and the addition of a hold-down clip at the rear bonnet sides. In the background are two CMP 7-ton drop-side semi-trailers and several 3-ton GS trucks. The C60S semi-trailer tractor appears to be fitted with 8.25-20 tyres on split rims.

Greasing registration 161902 from a trailer-mounted Tecalemit pressurised air, oil and grease delivery system. The engine-powered air compressor and operator can be seen behind the drums of oil and grease. The system was mounted on a sled with a canvas canopy and frame and was fully self-contained.

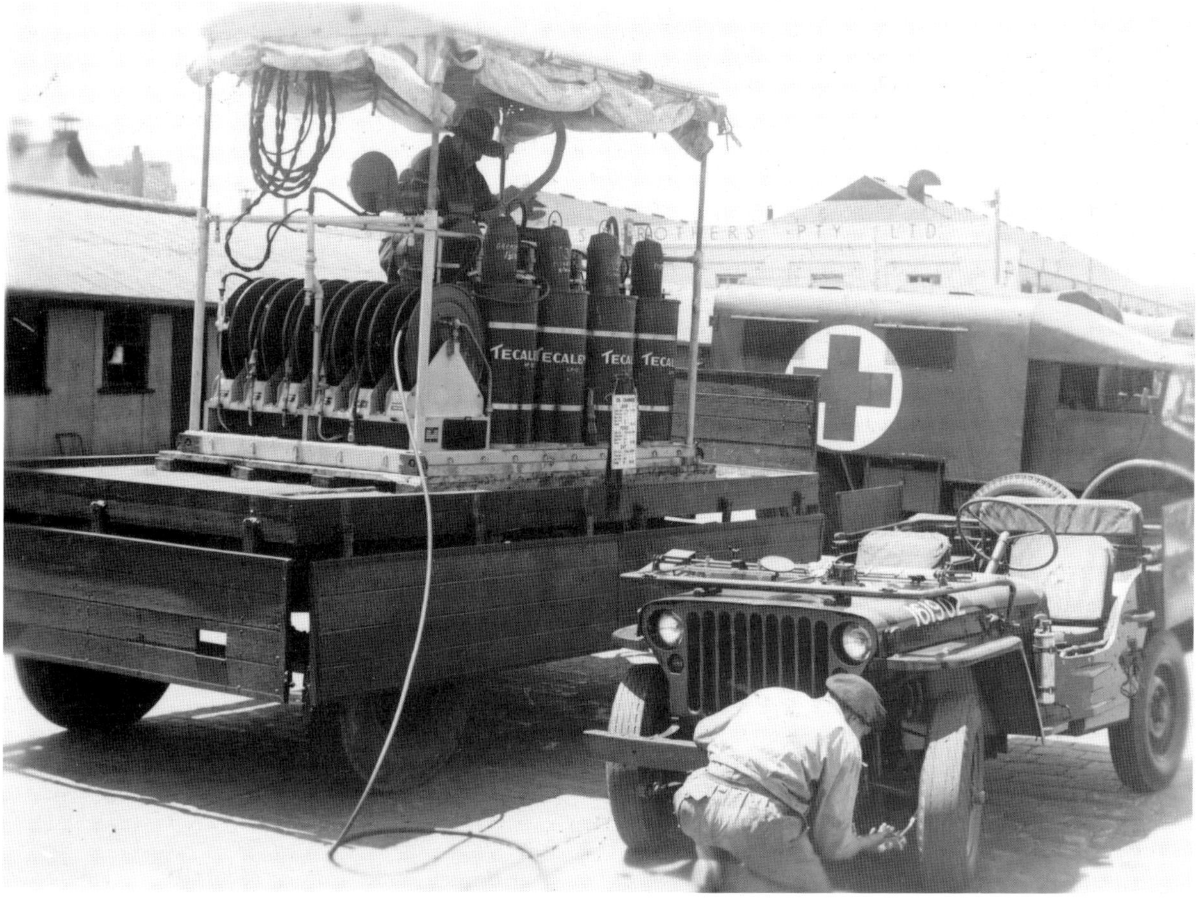

The sled could be carried on any suitable trailer or GS truck. A self-retracting hose drum for the air line and each type of grease and oil were mounted across the rear of the sled. Note the sign board detailing the types and quantities of oils used in jeeps, Ford and Chevrolet trucks. A 'Truck, 3-ton, Ambulance (Aust) No.2' can be seen in the background.

A different kind of maintenance: wiping away the moisture from the inside of the distributor cap of a mud-spattered Ford GPW while on a field exercise. Note the additional hold-down clip attached to the rear side of the bonnet, and the attachment of the ground strap to the second bolt of the bonnet hinge. The underside of the bonnet is fitted with a servicing diagram holder.

The field commissioners of the Salvation Army were renowned for their generosity and willingness to provide additional luxuries for front line troops. Major Jack Semmens is seen here dispensing hot drinks from urns perched on the front bumper and providing entertainment from a portable Zenith radio set. His jeep, a Willys MB registration 134959, has been modified with a solid roof, doors and side curtains. Additional catches have been installed on the rear edges of the bonnet to reduce vibration. The leading edge of the roof has the BCOF formation sign which remained on vehicles of the British Forces in Korea (BCFK) and the unit sign '68' of 66th Australian Infantry Battalion, 34th Australian Infantry Brigade, BCOF. The Salvation Army sign on the lower windscreen frame has each word cut out from a single piece of sheet metal which has then been fastened to a backing piece of sheet steel. The backing piece has then been glued to the lower windscreen frame. The rear vision mirror has been changed from the round, flat mirror to the rectangular type with a convex mirror usually seen on Canadian Military Pattern vehicles.

An Australian driver from 3 RAR about to cross a river in Korea while in the background, US Army engineers work on repairing the bridge. The jeep registration number does not fall within any of the Australian numbering sequences, so how the vehicle was obtained is open to question! Note the small, neat Tare weight painted below the axe stowage bracket, and the large roundel carried by British Commonwealth vehicles in Korea for ground and air recognition. The tyre pressure of 30 psi is painted on the bodywork above the centreline of each wheel. This is an early production vehicle as it is not fitted with a blackout driving lamp and guard on the front left mudguard.

The jeep remained the Army's light tactical truck in the post-war period. It was supplemented by a small purchase of Austin Champ trucks from Britain in the mid-1950s, but remained in service until the ¼ and ¾ ton Land Rover Series 2 vehicles were accepted for service in 1959. The remaining jeeps were then progressively withdrawn as the Land Rovers were delivered, with the last jeeps sent for disposal in the early 1960s. The jeeps pictured here are all on charge to the Eastern Command School of Technical Training of the Australian Regular Army indicated by the unit sign '103' painted onto the lower windscreen frame of several. The students and instructors were conducting an extended driving course in the Burragorang Valley area and along the banks of the Wollondilly River about 110 kilometres west-south-west of Sydney during 1950.

The jeeps show a wide variation in the way the registration number has been applied, both in size and position. Equally, the application of the 'Caution Left Hand Drive' sign shows significant variation, from letters almost 90-mm high to as small as 40-mm, and with the word 'caution' omitted in some, but included with others as an underlined word followed by an exclamation mark. Although the position of the bridge sign on the right side of the lower windscreen frame shows consistency, the size and form of the bridge weight '2' varies. All the jeeps appear to be finished in the late war satin finish Khaki Green No.3 paint rather than the post-war full gloss Deep Bronze Green.

A Ford GPA amphibious jeep on display at the MGO Branch at Wesley College, Melbourne in September1943. The vehicle was introduced into service as a light amphibious reconnaissance vehicle in areas where creeks, small bays and other stretches of water would have to be crossed during operations.

By October 1943, 161 amphibious jeeps had arrived in Australia with most in process of assembly at the contractor's works in Sydney, NSW and Geelong, Victoria. The limited Australian trials that had been carried out during September indicated serious shortcomings in the performance of the GPA, so further shipments were cancelled.

A newly assembled Ford GPA towing a 'Trailer, 8-cwt (Aust) No.3' during testing at Corio Bay, near the Ford plant at Geelong, Victoria. The GPA has yet to be assigned its Australian registration number. Note the sandbags around the trailer, used to simulate a full load. The GPA's poor performance particularly in surf or choppy conditions contributed to the decision to severely restrict their issue and use.

CHAPTER 3

Ambulances

The problem of clearing casualties from the front lines to the nearest aid station, and then further back to a Casualty Clearing Station became far more acute in jungle operations. Where tracks existed, they were often just foot paths through the jungle or kunai grass. The larger capacity military ambulances were simply too large, too heavy and too conspicuous to get anywhere near the front line. Casualties either had to be walked out assisted by a single soldier or carried out by four or more. It was hard on the casualty and temporarily reduced the manpower for fighting. The nimble jeep was soon being utilised to carry supplies forward and casualties back. With the value of a jeep as a makeshift ambulance quickly established, improvements were devised to improve the jeep's capacity for carrying stretcher cases. Ambulance conversions capable of carrying two, three or four stretcher cases were developed both at unit level and officially as an installation kit. They were highly successful and contributed greatly to the speed and comfort with which casualties were recovered from forward areas and hence to the survival rate.

Clearing casualties from forward areas beyond the reach of 30-cwt, four-berth ambulances was originally accomplished by two to four men carrying the stretcher to the unit aid station. This was slow and required considerable manpower. With the introduction of the jeep into service, soldiers soon found ways to carry stretchers. This shortened the time to get a wounded soldier to life-saving medical care. The simplest method of transport was to support the stretcher on the spare wheel and the lower windscreen frame, as seen here in Papua New Guinea. It required no alterations to the vehicle, but could only carry one stretcher at a time. Note the large '2' in a circular bridge sign on the lower windscreen frame. The jeep is a Willys MB, registration 79805, chassis number MB131536. It was disposed of to Dominion Motors immediately post-war.

Loading a wounded Australian tank driver onto an unmodified jeep on Labuan Island, North Borneo in 1945. As the stretcher was not secured in any way, safe carriage on an unmodified jeep was possible only over short distances on reasonably level tracks.

Multiple stretchers could be placed laterally over the bonnet and the canopy frame behind the driver on an unmodified jeep. However, this was not practical in forward areas due to the rough and narrow tracks. The two casualties across the back of the jeep experienced a particularly rough ride and all the stretchers snagged on the vegetation along the edges of the track. These deficiencies dictated that stretchers had to be mounted longitudinally and properly secured. Note that this jeep is fitted with dual rear wheels. This modification was an American invention by units in the Buna-Sanananda area of Papua New Guinea in early 1943 that also saw very limited adoption by Australian units. The modification used eight spacers 152-mm long x 38-mm of 9.4-mm mild steel plate welded inside the wheel rim to join to an outer wheel. The outer wheel had the centre section cut away to provide access to the wheel nuts.

The first official version of an ambulance jeep conversion was the 'Gear, Supporting, Stretcher, Light (Aust)' which could carry two stretchers. The installation of the supporting frames required only minimal modification to the jeep, but with the stretchers in position, the spare wheel, spare wheel mounting bracket and fuel can had to be dismounted and stowed on the rear floor of the jeep. There were two tubular stretcher mounting frames. The forward frame utilised the canopy frame stowage bracket. This frame had an angle steel strut which sat on the top edge of the two side panels and had a 12.7-mm rod which, when inserted, locked the feet of the stretchers and prevented them from bouncing off the frame. The rear frame was a large square made of steel tube. The lower tube was of a diameter slightly smaller than the aperture of the closed towing pintle. The frame was held in place by the closed pintle hook and a lug welded to the steel tube either side of the pintle hook. The weight of the frame and stretchers was supported by the pintle hook and bumperettes. The upper section of the frame had an angle steel piece which supported the feet of the two stretchers and incorporated a steel locking rod to hold the stretchers in position. The casualties were to be placed with their heads towards the front to provide some protection from the weather, as the canvas canopy did not cover most of the length of the stretcher. Although only capable of carrying two stretcher cases and one seated casualty, the main advantage of the 'Gear, Supporting, Stretcher, Light (Aust)' was that it could be very quickly installed or removed.

When not carrying stretchers, the rear frame could be folded forward against the rear canopy frame and held by a strap. It was not possible to stow the canopy frame in the down position with the 'Gear, Supporting, Stretcher, Light (Aust)' installed. Registration 125596 was a Willys MB-BRT, chassis number 197785.

The use of the two-stretcher supporting gear continued into the post-war period. This Ford GPW, registration 160457 with chassis number GPW213207 has been repainted in gloss finish Deep Bronze Green. Note that the rear stretcher supporting frame is of different construction to the original and has been strengthened by the addition of bracing across each corner. In addition to the registration numbers sign-written on each side of the bonnet, the number has unusually been painted onto the front bumper as well. The sign on the side indicates the jeep has been filled with range fuel. A jeep crank handle is perched between the front chassis extensions, rather than in the normal and safer stowage position behind the rear seat.

A two-stretcher frame equipped Ford GPW, registration 131118, chassis number GPW162746 in service in Korea. The stretchers have been equipped with two wide webbing straps with sliding buckles to secure the casualties. The jeep has the large air recognition roundel on the bonnet with a smaller version on each side. There are 30 psi tyre pressure markings painted onto the bodywork above each wheel. The RMO, a captain, sits in the front passenger seat while a soldier armed with an Owen Machine Carbine rides 'shotgun' in the rear between the stretchers.

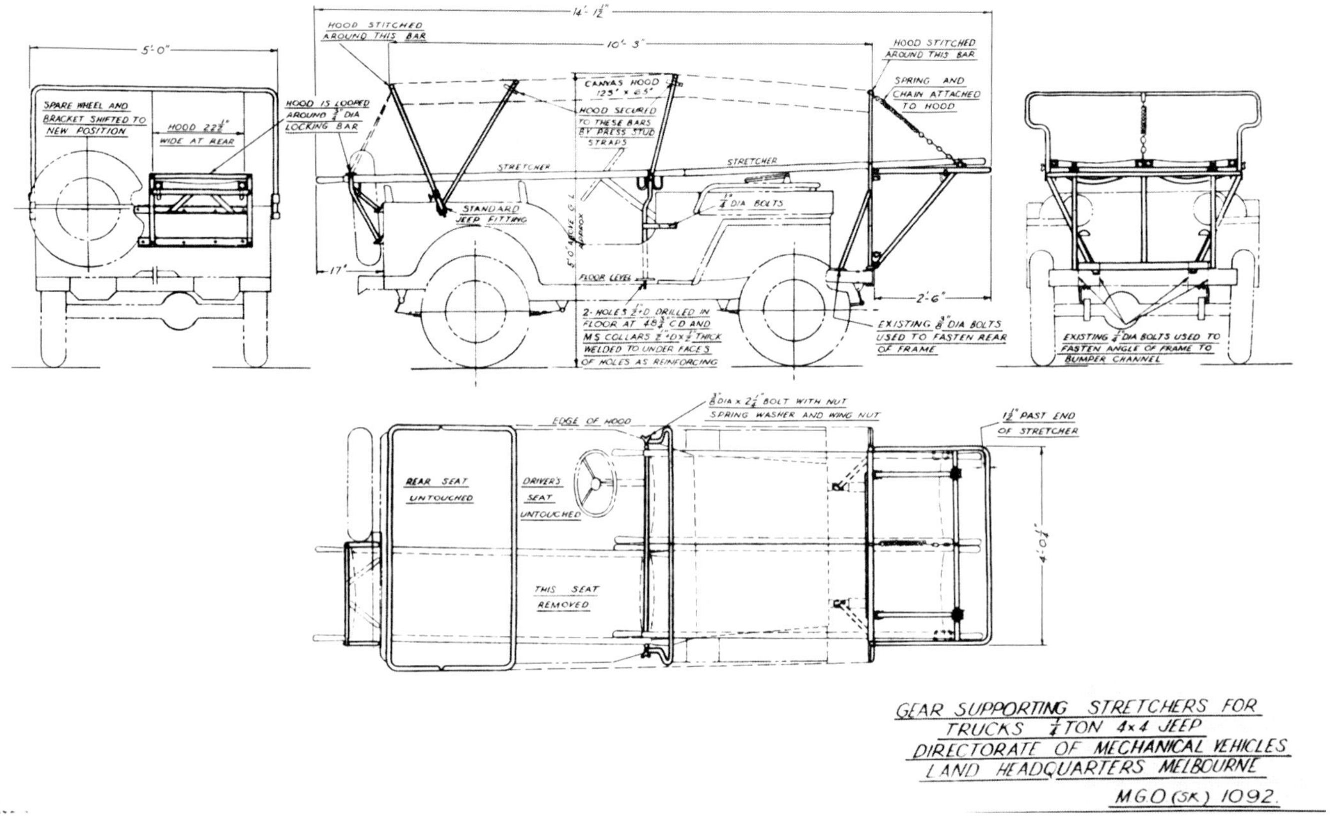

The 'Gear, Supporting, Stretcher' was developed to provide a capacity for the carriage of three stretchers and at least one seated casualty or a medical orderly, all protected from the weather by a canvas canopy. Modifications included removing the fuel can bracket from the rear panel and moving the spare wheel to that position, removal of the front passenger seat and the fitting of supporting brackets to the bodywork using 1/4-inch bolts. The stretchers protruded beyond the rear of the bodywork by 432-mm, and forward of the front bumper by 762-mm.

The 'Gear Supporting Stretcher' fully erected with three stretchers and the canvas canopy in position. This early iteration has the spare wheel mount removed and the spare fuel can bracket still in place.

Rather than a new, fitted single piece main canopy with a separate rear extension on the right side to cover the stretcher, this early trials version utilised the standard jeep canopy with a forward extension. Registration number 128732 was a Willys MB-BRT with chassis number MB227930.

Detail of the tubular stretcher support mounted to the front bumper and chassis extensions. The stretchers were prevented from bouncing off the supporting bracket by pins that were inserted through the feet of each stretcher. To prevent loss, the pins were attached to the support frame by short lengths of light chain.

Front view with the canopy folded to the rear and the three stretchers in position. The area behind the driver provided seating for a casualty or a medical orderly.

Rear view with the stretchers removed. The rear seat had to be folded down when carrying a stretcher in the rear right position.

A closer view, showing detail of the rear stretcher support. Note the lateral brace between the top of the rear mount and the fuel can strap bracket.

A side view of the 'Gear, Supporting, Stretcher'. The stretcher supports consisted on three tubular steel frames. The front frame was attached using existing bolts in the bumper bar and chassis frame. The middle frame required drilling holes in the floor and dash panel. The rear frame was attached by drilling eight holes in the back panel for the securing bolts.

Front view with the stretchers removed, showing the elaborate front supporting bracket which included a lateral brace shaped to accommodate the bow of the stretcher canvas when carrying casualties.

Detailed view of the supporting brackets attached to the floor and dash panel.

Detail view of the central and forward stretcher mounting frames. The swing-over bracket on the top of the central frame was to lock the arms of both forward stretchers against the frame. When closed, the bracket was secured with a pin which can be seen hanging by a short length of chain.

The RAAF's 75 Squadron independently arrived at a much simplified version of the three-stretcher ambulance jeep that utilised a continuos outer frame of steel tube supported at the front by a simple cross-braced frame attached to the front bumper and singular supports in the middle and at the rear. The carry handles of the two stretchers on the right side overlapped, so the frame was kinked upwards so the handles of the front stretcher were raised slightly so as not to interfere with those of the rear stretcher. A light square frame mounted to the front mudguards and braced to the front of the outer frame was used to support the front of an elongated canvas canopy with the standard jeep canopy bows providing the main support at the rear. Note that this design required the removal of the windscreen frame and the transfer of the spare wheel and mount to the left side of the rear panel. The squadron operated P40 Kittyhawk fighter aircraft and is famous for the air defence of Port Moresby during March-April 1942 and the Battle of Milne Bay during August-September 1942.

In an effort to load even more casualties aboard, units developed a single tier four stretcher version using three steel frames and the windscreen frame for support. The steel supporting frames were manufactured from angle steel, with the front frame attached to the front bumper and braced back to the top of each mudguard. This was high enough such that the two forward stretchers sat level between it and the lower windscreen frame with the windscreen in the open position. The two rear stretchers were supported at a similar height by a forward frame bolted to the side panels and an angled rear frame bolted to the rear panel and braced to the side panel. The bumperettes were removed. The spare tyre support was removed and the spare tyre carried within the rear stretcher support frame. This particular vehicle conversion was designed by Sergeant A. K. Sweetnam of 2/8th Aust Field Ambulance. The Ford GPW, registration 142930, chassis number GPW176648, was disposed of to the Royal Netherlands East Indies Army (RNEIA) at the end of the war.

The four-stretcher, single tier jeep ambulance is use. The rear supporting frames were covered in arc-mesh and used to carry the spare tyre and as additional stowage space. Owing to the height of the supporting frames, there was no need to remove the front passenger seat or fold the rear seat down. Hence, with only a single stretcher at the rear on either side, a casualty or medical orderly could be seated in the back of the jeep.

A casualty being attended to by a medical team while still aboard a jeep four-stretcher, single tier ambulance. Construction details of the rear stretcher supporting bracket can be clearly seen. The canopy has the embarkation details of the LSM number and the stowage position within the ship.

Medical officers attend wounded at a roadside casualty clearing station on Labuan Island, North Borneo, in 1945. The method of construction and mounting of the central stretcher supports of this single tier, four stretcher version can be clearly seen. The loop protruding forward of the support is to engage the leg of the stretcher to prevent lateral movement. Note also the location of the fire extinguisher on the rear slope of the front mudguard, and the axe carried in the lifting handle. An RAAF HAR-1 tipper from an airfield construction squadron can be seen passing by.

Another variation of the single tier-four stretcher jeep ambulance had a more elaborate angle steel frame with two cross pieces at the front with the stretcher legs held in place between them. The height of this frame was level with the lower windscreen frame. On this jeep, the inner windscreen frame and glass has been removed, but the outer frame retained to provide the rear support for the two front stretchers. The middle and rear support frames are much lower on this version, and the rear support frame much simpler in construction. It appears to utilise the bolt holes normally used for mounting the bumperettes, which have been removed. Because of the lower height of the frames for the rear stretchers, the spare wheel could not be carried within the rear frame, so this unit has devised a vertical mount between the stretchers.

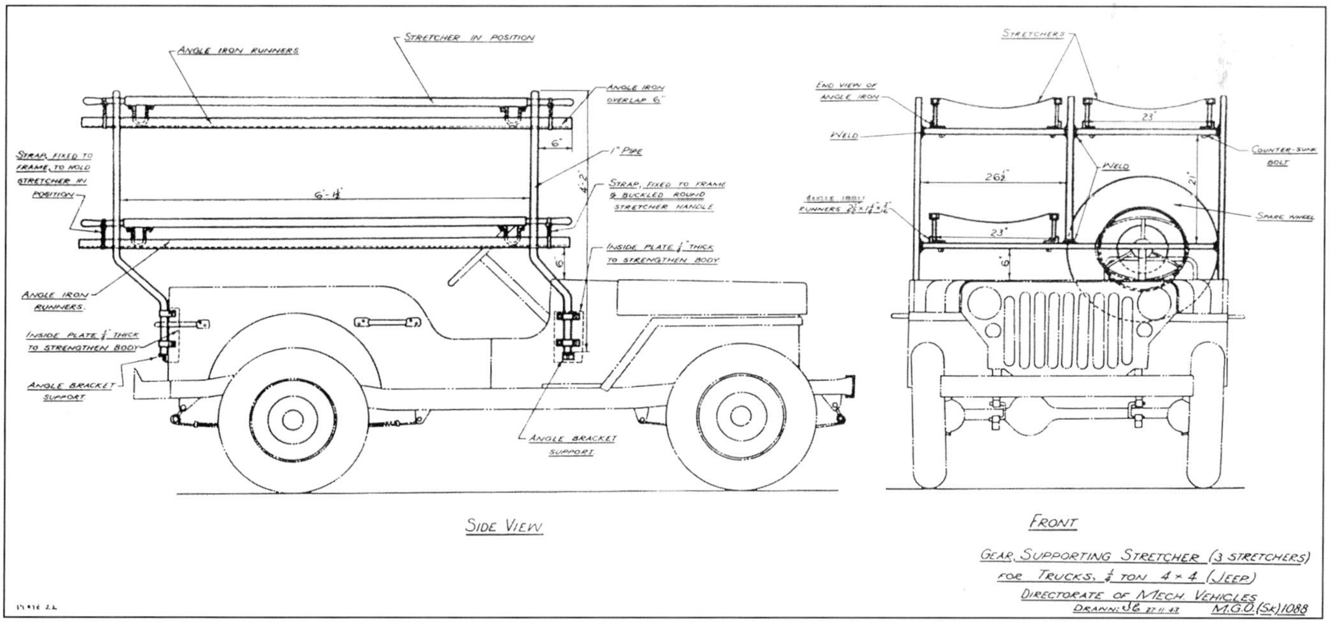

The final Australian stretcher-carrying conversion for a jeep was a two-tiered framework for carrying three stretchers. When originally devised, it was designated the 'Gear, Supporting, Stretchers (3 stretcher)'. However, once it was determined to issue the conversion as a complete kit for installation on a jeep at unit level, the nomenclature was changed to 'Kits, Gear Supporting Stretcher Light (Aust) No.3'. The kit consisted of all the parts and fastenings required for installation as well as installation instructions. The significant advantage of the two-tier conversion was that there was nothing interrupting the driver's forward vision, and only a minimal extension beyond the rear of the jeep.

'Miss Maureen', a Willys MB, registration 127620, chassis number MB196081, equipped with an early version of the two-tier, three-stretcher jeep ambulance conversion, transporting patients to a Casualty Clearing Station in the Ramu Valley, Papua New Guinea.

The early two-tier ambulance jeep used a simple tubular frame with locking bars to secure the handles of the stretchers. Unloading required the top-tier stretchers to be lifted up to clear the framework, then passed down with difficulty to other personnel at ground level. 'Miss Maureen', with the sub-unit designation B31, has a man standing on the scuttle to lift the stretchers. Note the folding locking bars on the upper frame.

Delivering casualties to an aid station set up in a native hut. This view of Willys MB-BRT registration 125679, chassis number MB188843, provides details of the rear mounting of the early two-tier jeep ambulance. An additional lateral brace was included on the left side of the frame rear to mount the spare wheel mounting bracket which had been removed from the right side of the jeep's back panel. The spare fuel can bracket has been removed, but the fuel can strap bracket was left in place.

Medical orderlies bring a casualty from the RAP to a waiting jeep ambulance for transport to a hospital. Aid posts were designed and equipped to stabilise casualties and provide short-term care for less seriously wounded. More severe cases were sent to a hospital further in the rear as soon as practicable. This is the early pattern of the two-tier stretcher supporting gear. Note the plate attached to the grille with the formation sign of 1st Australian Corps. The registration appears to be 136753, which was a Willys MB, chassis number MB231633.

The 'Kits, Gear Supporting Stretcher Light (Aust) No.3' was the final version of the two-tier, three-stretcher jeep ambulance conversion. Experience with the early versions had resulted in several improvements including manufacture from substantially heavier tubing, the provision of side rails and the use of angle steel longitudinal runners to slide the stretchers on and off the framework from the rear. This meant that stretchers did not need to be lifted clear of the framework, so casualties could be unloaded by orderlies at ground level. Contract CD-E9165 for the manufacture of 260 kits was awarded to Cowan and Sons in mid-1944. Registration 128459 was a Ford GPW, chassis number 65449. Note the canvas windscreen cover and the height of the spare wheel relative to the stretcher frame.

The two-tier stretcher supporting kits were also used by the RAAF. 42110 LAC Frank Kuffer, a sergeant medical orderly serving with No.10 Local Air Supply Unit on Bougainville in 1945, is seen in the unit's ambulance jeep. The spare wheel bracket has been mounted so that the wheel sits much lower than originally designed.

Heavily laden ambulance jeeps lined up ready to board an LSM. The 'Gear, Supporting, Stretcher Light (Aust) No.3' is the latter type with longitudinal runners to support the stretchers and allow the stretchers to be loaded or removed easily from the back of the vehicle. The jeeps are equipped for deep fording with an extension on the carburettor intake and on the exhaust pipe. The bonnet has been removed and replaced with a piece of canvas. The bonnet was stowed on the vehicle, ready to replace after landing. A steel tow rope has been attached to the front, with the running end in the passenger side of the jeep, ready to hand forward in case recovery during the landing is required. Chains are fitted to all wheels. The top tier is heavily loaded with stretchers and tent poles and a canvas cover has been fitted. The additional flat cross bracing at the front of the angle steel stretcher slides was a functional improvement that prevented the stretcher legs from being pushed too far forward. A sheet steel plate has been fixed to the radiator guard which has the jeep's details of the engine and chassis number. The chassis number, GPW189472, equates to registration number 142825, which was disposed of immediately post-war to the RNEIA.

Not long after the initial issue of the two-tier stretcher gear kit, operational experience showed that the structure was able to rock forward and backward when loaded and moving over uneven ground. This flexing caused the mounting bolts to tear holes in the jeep body and become loose. To overcome this, two additional angle braces were added between the rear vertical frame and the top edge of the jeep body adjacent to the front seats. Registration 128205 was a Ford GPW, chassis number GPW51282, which was retained in service until 1960, when it was written off by Board of Survey at 2nd BOD, Sydney, NSW. (See also Appendix 1-11)

CHAPTER 4

Communications

Although the Australian Army developed wireless vehicles on large and heavy chassis, these were difficult to land in the early stages of an amphibious operation and difficult to manoeuvre in jungle terrain. Yet the Army were slow to develop a kit for equipping a jeep as a wireless communications vehicle. So slow, in fact, that the prototype did not materialise until 1945. Although a quantity was subsequently ordered, the contract was cancelled at the end of the war. Hence it was left to units to cobble together mountings in jeeps for the various wireless sets on issue, mainly the No.11, No.19 and No.22 sets. The universal problem of operating any of these sets within a jeep was the charging of batteries. The jeep's electrical system was 6 Volt DC, whereas the various wireless sets operated on 12 Volts DC. Providing power to a wireless set from two 6 Volt batteries wired in series gave the option of recharging one battery at a time by connected to the jeep's electrical system, but this was very cumbersome and required running the jeep engine for long periods. A kit for converting dedicated wireless jeeps to 12 Volt DC electrical systems was briefly considered but abandoned in preference to carrying a separate, self powered charging set such as a Chore Horse.

Units also devised mountings that integrated all the components of a complete wireless station (except the batteries) into a single unit, often a timber box. This provided greater protection from external damage and the elements. It also made installation and removal less complicated.

Amphibious operations required the establishment of a powerful 'rear link' wireless set as soon as possible after landing. Initially, heavy wireless units had transferred the wireless station ashore in crates that then took a couple of days to set up and commence operations. To minimise the set-up time and increase unit mobility, heavy wireless units devised ways to carry the WS No.133 (Aust) and its necessary supporting equipment in as few as two jeeps and trailers. Set-up time from landing to establishing communications with higher headquarters was reduced to as little as a couple of hours.

Troopers from an armoured unit monitor a WS No.19 Mk.2 mounted in the rear right corner of this Willys Model MB, chassis number MB197863. The wireless set mounting and aerial brackets were manufactured by the unit's Light Aid Detachment. The WS No.19 is equipped with a canvas cover for weather protection. The aerial variometer and the switch box are mounted on the top of the set, with the aerial cable looped around the lifting handle before being joined to the bottom of the A-set aerial mount. The WS No.19 Mk.2 was the communications set commonly used in both armoured and support vehicles of armoured units. This jeep, which served with an armoured unit in Borneo late in the war, was transferred to Allied Land Forces South East Asia shortly after the cessation of hostilities.

A different mounting for a WS No.19 Mk.2 in the back of a 'Truck, ¼ ton (USA)', this time fully enclosed in a wooden case with the bank of batteries in boxes on the back seat. These were usually 6 Volt batteries wired in series to provide the necessary 12 Volt DC power supply. A morse code key can be seen adjacent to the variometer, while the Control Unit No.1 which allowed switching between the 'A' and 'B' sets, is sitting conveniently on the top of the case. The vehicle is serving with 2/4th Aust Armoured Regiment. As there was no officially standardised Australian design for mounting wireless sets in jeeps until the end of the war, units devised their own methods to achieve this.

Even the most robust of wireless sets will need repair from time to time. The jeep bonnet and front bumper make convenient work benches for a wireless technician as he isolates the fault using a 'Multitesters No.2 (Aust)' while his associate prepares a blow torch to heat an iron for soldering.

A jeep in Papua New Guinea equipped with a WS No.11 (Aust) in its carrying frame along with the Supply Unit LP while the Supply Unit HP is in the tray on top of the set. A Wireless Remote Control Unit 'A' is on the ground adjacent to the jeep. The WS No.11 (Aust) operated on 12 Volt DC, with power normally provided by two 6 Volt, 75 Ah batteries connected in series. Each battery was housed in a wooden box with carry handles. The jeep has had the windscreen frame removed, and has wheel chains fitted. The signage on the side of the jeep scuttle has been obliterated by the censor.

The pilot model Australian wireless and telegraphy jeep included a rear mounted cable drum carrier for laying and recovering signal wire. The original intention was to provide wireless jeep conversion kits from the USA which included changing the jeep electrical system to 12 Volts to enable charging of the wireless batteries, but this was abandoned in favour of an Australian-designed system which utilised a 12 Volt DC charger driven from a power take-off (PTO). The wireless mounting frame was to be universal to fit a range of sets including the WS No.11, WS No.19 and WS No.22 (Aust). Aerial mounts were to be within the width of the jeep to avoid damage in close country, and a 300-Watt auxiliary charging set was to be carried.

The design project commenced in early 1944, with the mock-up completed by late April. The requirement was later expanded to include the mounting of a frame to hold a standard cable drum of signal wire, and dispensed with the PTO with the auxiliary charging set providing the only means of wireless battery charging. A pilot model 'Kits, Wireless Telegraphy and Cable Laying' was manufactured and tested and order CD-E10675 placed in mid-1945 for 586 but with the end of the war, the order was cancelled. Registration 128010 was a Ford GPW with chassis number GPW66429. In 1945 during the build-up in Australia for the invasion of Japan, the jeep was transferred to the RAN for eventual transfer to the Royal Navy.

This jeep is fitted with a 'Frames, Cable Laying (Aust) No.1 Attachments MV' in the rear compartment for laying and retrieving signal cable. The frame fitted the standard steel cable drum and had a simple winder and brake to control the rate of release or retrieval. Attached vertically to the front bumper is a short length of steel tube which holds a wooden handle with an attachment for tracing along a cable to isolate breaks. The attachment has a loop of round rod with a small opening to allow a cable to be picked up and held while the soldier walks along the route of the cable. In this case, it is being used to pick up the cable clear of the front of the jeep while being retrieved. The unit sign of a 2-4 over 74 in red on a horizontally divided background colour of white over blue shows this jeep is serving with the signals detachment of 2/8th Aust Field Regiment. Note the 'Ropes, Towing, Steel, Medium (Aust) No.1' loosely wrapped around the front bumper and chassis extensions, and the P-59 Mk.1 25-pdr projectile boxes attached to each front mudguard to provide some extra stowage space.

Signallers running out communications cable using a 'Frames, Cable Laying (Aust) No.1 Attachments MV' and a cable-following handle during the early 1950s. The jeep is an early Willys MB sourced from the Royal Navy, most probably from British stocks being assembled in Australia for the invasion of Japan. When the war ended abruptly in 1945, many of these stocks were transferred to the Australian government. The jeep remained in Australian Army service until the 1950s. It has seen heavy usage judging by the bends and dents in the rear panel. Note that the rear reflectors appear to have been overpainted. The unit sign attached to the rear right bumperette is white over blue, indicating a signals unit.

The same Willys MB, registration 167157, this time equipped with a wireless set mounted on the rear left sponson and a battery box behind the driver's seat. The vehicle has been painted in gloss Deep Bronze Green. Various fittings such as the canopy bows and the windscreen hold-down clamps are missing, but it does have additional bonnet hold-down clamps fitted on either side at the rear of the bonnet to minimise vibration. A white over blue unit sign of a signals unit is painted onto the left side of the lower windscreen frame.

In order to adapt ¼ ton trucks and trailers to the needs of specialist units, it was not unusual to have several vehicles operate together, with the loading of no one vehicle or trailer able to carry out the unit's functions without the others. Detachments, Wireless Section (Heavy) operated the WS No.133 (Aust), the Australian version of the British WS No.53. This was a medium power, long range transmitter designed primarily for static and semi-mobile installations. The complete unit was bulky, cumbersome, and heavy. Initially in Australian amphibious operations, the Detachment's equipment had been landed on the beach in crates which then had to be manually handled to a suitable location, unpacked and assembled before the station could come on the air – a process that could take as long as two days. To streamline this process and provide a degree of mobility to the Detachment for the invasion of Tarakan Island on 1 May 1945, 1st Australian Corps managed to load two ¼ ton trucks and trailers with the Detachment's entire equipment. The WS No.133 (Aust) was carried across the rear compartment of a ¼ ton truck, with the main rectifier bolted to the passenger side floor (after removing the seat). Although used successfully during that operation, later testing by 1st Australian Independent Signal Group found that both the ¼ ton trucks and trailers were grossly overloaded by as much as 91%. This was unacceptable to the Director of Mechanical Vehicles. The formation sign on the jeep lower windscreen frame is the bulldog over a boomerang in white on a black background of 1st Australian Corps. The shipping details on the side of the jeep are unusually detailed. While the length, width and height differ slightly from the official measurements, the list also includes the cubic weight (CW), that is, the dimensional or 'dim' weight, with the canopy up or down. The Unit Serial Number signs on the jeep and trailer are at odds. The sign on the jeep is 45677 of 13th Aust Wireless Section (Heavy), while the trailer has 61340 of 1st Aust Wireless Section (Heavy). Both units were serving with 1st Aust Corps at the time of the landings on Tarakan.

During amphibious operations in the New Guinea and New Britain areas, heavy wireless detachments of First Australian Army trialled the carriage of a complete heavy wireless detachment in three jeeps and trailers. This resulted in a maximum overload of 25% which was acceptable in the circumstances, particularly as the wireless station was operational only 1 hour 22 minutes after landing. In this configuration, the WS No.133 (Aust) was carried in an 8-cwt trailer while the duty signaller had a desk in the back of a jeep.

The duty signaller sat at a desk in the first jeep with two Reception Sets No.4 (Aust) mounted on the sponson to his right with the Loudspeaker, 5-inch No.2 (Aust) on top. Other items carried in this jeep were a battery charger, six 6 Volt, 75-AH batteries, a Telephone Set 'L' (Aust) No.2, a remote control unit, aerial sets, spare valve boxes and sundry smaller items of equipment.

The first jeep towed a trailer equipped with a Howard 4.5 KVA, 240 Volt Petrol Electric Generator, a power extension cable, six 'Containers, 4-gallon, Cylindrical' (five petrol, one with oil), two 'Pourers, Petrol, Flexible' and a tool kit for the generator. The trailer was modified with adjustable rear corner supports which folded up toward the centreline when stowed, and an arched tubular canopy frame. The fitted canvas canopy could be either removed or partially laid back when the generator was operating. Registration 156398 was a US pattern trailer, frame number 45167.

The second jeep towed the trailer with the WS No.133 (Aust) transmitter, a Wavemeter, headphones, a Telephone Set 'L' (Aust) No.2, a remote control unit, a case of spare wireless parts, a Multi-meter (Aust) No.1, and sundry small items of equipment. The trailer was modified with an arched tubular canopy frame and fitted canvas canopy. The frame had a hinged section on the left side which could be propped open to allow access to the wireless set. Each rear corner of the trailer was fitted with an adjustable support.

The jeep towing vehicle was fitted with brackets to carry an 'Apparatus, Cable Laying, No.3 (Aust)', six cable drums – two of single D3 and four with twisted (double) D3 cable, a 6 Volt DC, 75 amp-hour battery and a jar each of electrolyte and distilled water. When setting up as seen here, the jeep detached the trailer carrying the WS No.133 (Aust) and moved away while laying the remote control cable to the jeep with the duty signaller.

The third jeep was unmodified. It carried the rations and personal kit of the detachment. It towed the trailer seen here which was equipped with a Howard 4.5 KVA, 240 Volt Petrol Electric Generator, plus two sets of aerial poles, stay plates, spikes, aerial ropes and a sledge hammer. The aerial pole sets were stowed within racks each side of the centrally-mounted generator unit. Trailer registered 156384 was a US pattern trailer with frame number 45988, which was disposed of in New Guinea soon after the end of the war.

CHAPTER 5

Gun Towing

The main anti-tank and field guns used by the Australian Army during the period of the jeep's service were the 2-pdr, 6-pdr and 17-pdr Tank Attack guns, and the 25-pdr and 25-pdr Short field guns. The 17-pdr Tank Attack gun and the 25-pdr field gun were simply far too heavy for towing by a jeep on operations. The use of a jeep as an artillery tractor was tried with the 2 and 6-pdr tank attack guns, and the 25-pdr Short field gun with varying success. In a report on operations in New Guinea by 2 Aust Corps for the period 8 October 1943 to 1 March 1944, it was concluded that, due to the excessive weight, the jeep was not a satisfactory prime mover for the 2-pdr Tank Attack gun. Nevertheless, jeeps were trialled as tractors for the heavier 6-pdr Tank Attack gun and the 25-pdr Short field gun, though not always successfully. Simply, an unladen or lightly-laden jeep could tow the 25-pdr Short field gun, but it took several jeeps and trailers to move the gun and all the paraphernalia required to operate it such as camouflage nets, picks, shovels, cleaning equipment, and first-line ammunition. It was much more economical to use the larger artillery tractors in most instances.

Jungle warfare dictated the requirement for a much lighter and more easily moved field artillery weapon than the normal 25-pdr field gun. Readily available alternatives did not deliver the same impact on target as the 25-pdr field gun, so in late 1942, a portable or pack version of the 25-pdr known as the 25-pdr Short was developed.

Initial firing trials of the prototype were conducted at Fort Gellibrand, Victoria, on 5 December, after which the gun was transferred to MEE at Monegeetta for towing trials with a jeep. These images show the prototype 25-pdr Short equipped with the large ordnance weight bolted to the cylinder block to reduce the gun's bounce upon firing. The towing tests showed that the gun spade was much too low as it ploughed the ground when crossing shallow ditches and depressions. As a trial, the pintle hook was raised 280-mm, but this did not overcome the problem of the gun carriage making contact with the jeep bodywork in tight turns.

Towing trials with a Willys MB, registration 76969, chassis number MB138940, towing a production 25-pdr Short field gun over undulating ground at MEE Monegeetta, Victoria. The gun carriage is equipped with a goose-neck trail eye in an attempt to overcome the problem of the trail spade digging into the ground. The jeep pintle hook was 483-mm, which resulted in the trail spade having ground clearance of only 298-mm when the goose neck was in the down position. The jeep was disposed of to the Post-Master General's department soon after the end of the war.

Despite the additional ground clearance afforded by the goose-neck, the trail spade still ploughed the edge of a shallow ditch during cross-country towing trials.

With the jeep steering at full lock, the gun carriage made an angle of 38-degrees from the jeep centre-line, which resulted in the trail spade being very close to making contact with the bumperette which has been attached to the jeep upside down. Note the additional holes around the rear wheel arch, the result of the jeep being used for several different trials and functional improvement tests.

Some artillery units devised their own modification to overcome the trail spade making contact with the ground. This unit, either 2/2nd or 2/3rd Aust Field Regiment, 6th Australian Infantry Division, attached a vertical post to the rear of the jeep with the pintle hook relocated to about 812-mm above the ground with the jeep unladen. The post stood away from the rear of the body by about 152-mm. It was attached to the pintle hook's original mounting point at the bottom, and had a short horizontal strut which attached to the top edge of the jeep rear panel. The positioning severely restricted the jeep's turning circle, particularly to the right, as the trail spade made contact with the spare wheel.

The 25-pdr Short field guns of 2/7th Aust Field Regiment being towed past the saluting base during a parade at Mount Garnet, Queensland, in July 1944. The jeeps are unladen. While the jeep was able to successfully tow the 1,320-kg 25-pdr Short field gun on good roads, great care had to be exercised when towing cross-country due to the lack of ground clearance of the trail spade and the extremities of the gun's trail coming into contact with the rear of the jeep. The problem was exacerbated with the jeep heavily laden with gun stores or ammunition. Note the spare fuel cans carried on these jeeps are standard, rectangular 'Cans, 2-Gallon' rather than the larger 5-US gallon jerrycan.

Towing the 25-pdr Short field gun presented problems because the track width was wider by 250-mm than the jeep towing vehicle. In soft ground, the tractive effort required was markedly increased, and in close country, a gap just wide enough for the jeep had to be widened for the gun to pass. Consequently, experiments were conducted with a gun with a track width reduced to the same as the jeep towing vehicle. While successful, the project was discontinued on 11 August 1945.

An important aspect of the reduction in the gun's track width to that of the jeep was embarking onto and disembarking from an ALC-20 landing craft. Four guns and two jeeps could be carried but embarking and disembarking was much faster with the jeeps and guns of the same track width.

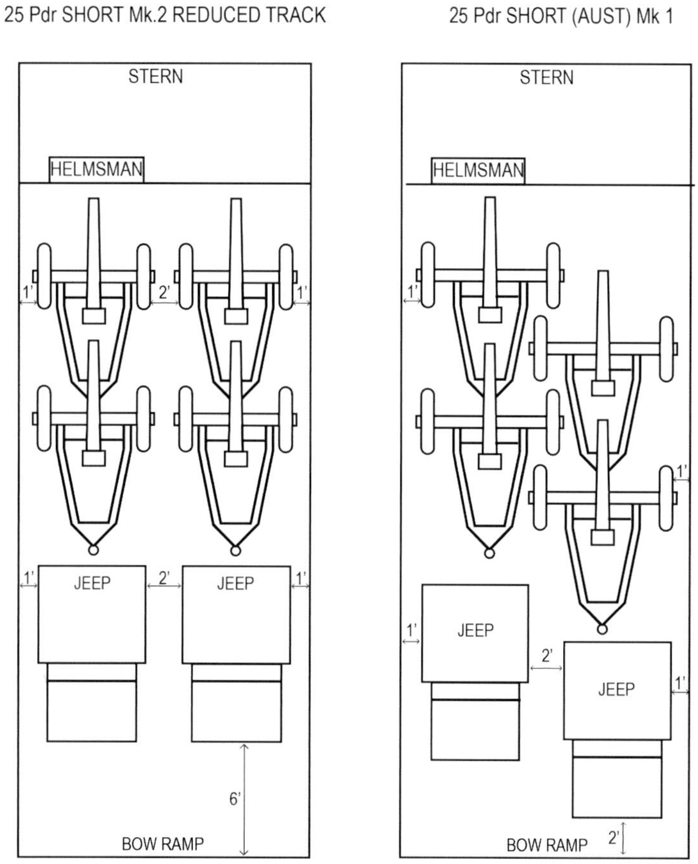

This diagram illustrates the reasons why the equal track width jeep and gun combination was faster to load and unload and could be carried out in any sequence, with both jeeps able to leave the ramp simultaneously if needed. With the wider gun, the overlap of the wheels meant the guns could only be loaded and unloaded singularly and in sequence.

Some preliminary trials under Project P2/6 were conducted in late 1944 to see if a single jeep could be successfully used as a tractor for the 6-pdr Tank Attack gun. The conclusion was that a jeep was a capable tractor but only for short periods. However, the Director of Mechanical Vehicles ordered that the jeep must only be used for this purpose in exceptional operational circumstances. This then led to the development of a tandem hitch to link two jeeps together to provide a low silhouette tractor capable of traversing the full range of terrain when towing a 6-pdr Tank Attack gun. Trials were conducted at MEE Monegeetta in November 1944. In addition to the 1,064-kg weight of the gun, the jeeps were weighted with local pattern carrier track to simulate the approximately 1,000-kg weight of gun stores, first-line ammunition and crew that would need to be carried on the jeeps. The overall length was 12.04-metres and total weight was 4,051-kgs.

The trials emphasised the low silhouette the jeep offered when moving the tank attack gun across country. The overall height to the top of the driver's head was only 1,473-mm, and to the top of the jeep's spare wheel, only 1,372-mm.

The low height allowed easy concealment in underbrush or jungle terrain. This was seen as a distinct advantage when placing a tank attack gun into an ambush position with its towing tractors concealed as near as possible to the gun position. The advantage of concealing the tractors close by was that, once the position was compromised by the first shot, the gun could be recovered without delay.

The length of the tandem hitch combined with its built-in spring extension provided an adequate maximum height difference when traversing uneven terrain. The jeep tractors were able to pass over the standard ground clearance knife edge used in the testing of all potential Australian Army vehicles. The hitch utilised the pintle hook on the leading jeep, and a special hitch attached to the front bumper bar with a brace to each chassis side rail on the following jeep.

The length of the hitch allowed lateral movement of the leading jeep up to 425-mm either side of the centreline. Note the coils of MG Carrier track to simulate a full load of stores.

Sharp turns were possible but at full lock, the front bumper of the following jeep made contact with the bumperette of the lead jeep. Nevertheless, the tandem hitch was considered suitable for use in difficult terrain at slow speeds. The jeeps had ample power and traction and offered distinct advantages in terms of concealment. On roads at higher speeds, however, the tandem hitch was a liability. Drivers had to be coordinated and alert at all times, especially when braking. Emergency stopping by the lead jeep inevitably caused the vehicles to jack-knife. When the following jeep accelerated sharply it pushed the lead jeep sideways. At the beginning of 1945, with the negative aspects of the design outweighing the advantages, it was decided not to introduce the tandem hitch into service.

During the 1950s, the 6-pdr Anti-Tank gun remained as a deployable first-line weapon in time of war. The jeep was used as a gun tractor, but only for displays and parade purposes, as seen here in service with 2nd/14th Light Horse Regiment (Queensland Mounted Rifles). The jeep registered 130330 was a Ford GPW with chassis number GPW119714. It was written off from 1st BOD, Queensland, in November 1959. Note the line of bolts across the face of the lower windscreen frame. These secured a remote turn signal arm to the inside of the frame which was activated by the driver when preparing for a turn or stop. Before the widespread introduction of indicator lights, hand signals were used from the right side of a vehicle to indicate intentions to the drivers behind. The driver seated on the left therefore required a remotely controlled signal arm to indicate from the right side. The remote signalling arm was common to all left hand drive military vehicles in use on Australian roads in the post-war period.

Definitely only for show! These jeeps, finished in gloss Deep Bronze Green paint, have been stripped of their windscreen frames, canopy frames and most of the supports so they can be used as tractors for 25-pdr field guns. The guns weigh about 1,500-kgs without the firing platform which the jeep is capable of towing, but only on reasonably level ground for displays and parades. Note the difference between the application of the registration numbers in terms of font, size and position on the bonnet sides.

The early development of the 'Kits Recovery' was much more elaborate than the final version issued as a kit. The winch was placed to the right of centre with an anchoring point in the centre of the bumper bar, the intention being to provide a pulley block on the winch cable and attach the running end to the anchor point. This would have provided a 2:1 winch ratio, which, while requiring twice as much winding by the operator, was less strenuous.

The winch was mounted so that it projected beyond the bumper. To protect the winch, additional bumper blocks were attached to either side which still allowed use of the issue crank handle. Note the lateral mounting strut between the chassis members and the angled braces. The winch had a detachable extension handle that was carried inside the jeep and inserted into the winch when required. It extended past the side of the jeep to allow operation without striking any part of the vehicle.

The issue 'Kit Recovery' mounted the winch on the centreline of the jeep, and dispensed with the pulley block, anchor point and bumper extensions, but retained the lateral mounting strut and angled braces. The bumper extensions were not required because the winch was mounted further back, necessitating the cutting of a radiator guard bar at the bottom and bending it to the side in order to clear the large winch intermediate gear. With the winch centred, the winding handle required an intermediate support strut made from angle steel which was bolted to the outside face of the bumper bar. A bearing attached to the strut supported the shaft of the winding handle. The shaft was fixed in position and had a hinge to allow the winding handle to be folded back and held by a clip attached to the front right mudguard when not in use. Contract CD-E8745 for the manufacture of 230 'Kits, Recovery' was awarded to the Ford Motor Company of Australia (FMC) in mid-1944. By 1 July, the pilot model had been delivered. Two further contracts were awarded to Ford, the last being CD-E10191 in mid-1945 for the manufacture of 300 'Kits, Recovery', but delivery was held up due to the wire rope being unavailable. In August 1945, the undelivered portions of the contracts amounting to 650 kits, were cancelled.

Several special purpose conversions of jeeps required a source of power with the vehicle stationary. To achieve this, Marbutt Engineering Pty Ltd in Brunswick, Victoria, were contracted in early 1944 to manufacture a pilot model PTO and governor under project V229. The result was a PTO assembly mounted on the rear of the transfer case with a shaft and sliding gear connection to the main shaft from the transmission. The PTO was engaged using a lever mounted on the floor of the jeep. A drive shaft connected from the PTO to a main bearing assembly attached to the underside of the centre frame crossmember. This large main bearing supported a short shaft with a pulley for multiple V-belts, with the number of pulley grooves and the diameter of the pulley dependent upon the application. The application in this image is for the 12 and 30 Volt generators of the Starting and Charging conversion, with each generator driven by three belts. To maintain adequate power from the jeep engine as the load from auxiliary equipment changed, a governor was mounted onto the left side of the engine and driven by a v-belt from the crankshaft pulley.

The 'Kits, Welding' utilised the Marbutt Engineering Pty Ltd PTO unit to drive an electrical generator mounted in the rear compartment, along with a welder control board and welding leads. The pilot model was tested and accepted for service during February-March 1944. There were two supply contracts: CD-E8825 issued in mid-1944 was for 94 kits, and CD-E10113 issued in early 1945 was for 84 kits, at a cost of £96 each. The 'Kits, Generator, 5 KVA' was very similar in layout to the welder. It utilised the same PTO but with a different pulley to power a 5 KVA, 415 Volt, 50 cycle three-phase generator. The electrical components were the same as those used for the portable 'Generating Set, Alternating Current, Petrol Electric, 5 KVA' already on issue. The control board was wired so that both 415 Volts, three-phase and 240-Volts, single-phase power was available. The kit was designed to provide power and light for workshops operating in jungle conditions. The prototype was approved on 7 June 1944. Subsequently, two supply contracts were issued to Marbutt Engineering Pty Ltd: CD-E8831 was for 60 kits, and CD-E10113 was for 84 kits. Both the welder and generator kits were issued to workshops and Light Aid Detachments up to Brigade level. The 'Kit, Welder' shown here is a restored example. The jeep is a Ford GPW, registration 155133, chassis number GPW164870 which was retained in service until early 1960. Markings are 2-4 over 54 in white on a horizontally divided background of blue over yellow over red which is for the 2/4th Aust Armoured Regiment Workshop. The formation sign is the crocodile and palm tree over a boomerang in white on a black background for the 4th Australian Armoured Brigade Group.

Detail of the Clayton Joel and Company welder kit control board. The three V-belt drive pulley is visible behind the board to the left. The board serial number is 47, so was supplied in 1944 under CD-E8825.

The generator and rear of the welder control board. The welding leads can be seen coiled on the floor. Adjacent to the driver's seat is the PTO engagement lever.

Cpl V. Jessup strikes the arc on a piece of steel held in a bench vice while undertaking equipment repair at 110th Aust Brigade Workshop Type B, (AIF), 6th Australian Infantry Division at Cape Wom, Papua New Guinea on 29 August 1945. The jeep behind him is equipped with a 'Kits, Welding'.

A jeep equipped with a 'Kits, Welding' parked under an awning at 2/56th Aust Light Aid Detachment, 7th Australian Infantry Division, at Kairi, Queensland in March 1945.

In the post-war period, a RAEME unit reconfigured a PTO-equipped jeep in the late 1950s by substituting an air compressor for the front passenger seat, placing the welder on the passenger side of the rear compartment, adding a stowage box and a stand for the compressor hose and welding cables. The jeep is finished in gloss Deep Bronze Green paint with the registration stencilled in white. The jeep had originally been supplied to Australia as a reconditioned vehicle under the Lend-Lease scheme. It was a Ford GPW, chassis number GPW79235 and when it arrived in Australia, it was equipped with a Willys engine, number MB113863. The jeep was written off by Detachment, 3rd BOD Seymour, Victoria in December 1960. Note the remote control signalling arm attached to right side of the lower windscreen frame.

Craftsman Ron Green operates a lathe mounted in the back of a jeep converted to a 'Truck, ¼ ton, Machinery'. It is equipped with a lath, drill, grinder and battery charger, with electrical power provided by a generator mounted on the floor and powered by a PTO. There were two basic variants of the machinery jeep developed in late 1943 by the MGO Experimental Workshop, each with a different suite of equipment. This jeep is part of the equipment of 2/4th Aust Armoured Regiment Workshop co-located with C Squadron, 2/4th Aust Armoured Regiment of 4th Aust Armoured Brigade Group at Aitape, Papua New Guinea, in February 1945. The formation sign of 4th Aust Armoured Brigade Group, a crocodile and palm tree over a boomerang, in white on a black background, has been stencilled twice on the rear of the jeep. The unit sign of 2-4 over 54, on horizontally-divided background colours of blue over yellow over red appears on the left bumperette, while the sub-unit sign of a 'C' is painted on the right bumperette. The faintly visible registration appears to be 128295 which was a Willy MB, chassis number MB197806.

Units were quick to improvise different attachments and modifications to the basic jeep. A workshop unit of 4th Australian Armoured Brigade Group in Papua New Guinea modified a jeep as a small mobile jib crane. The base of the jib used the same method of support and prevention of lateral movement as the two-stretcher ambulance, that is, a tubular horizontal member latched within the jeep pintle hook, and supported by the bumperettes. At each end was a simple hinge which attached to each end of the tubular members of the A-frame jib. When erected, the jib members were joined at the apex by a bolt. A cable from the apex was secured to a bracket anchored through the jeep floor to the machine gun mounting plate. A standard chain block hoist was hooked onto the apex of the job for lifting. When stowed, the apex was unbolted and the individual arms of the A-frame laid along each side of the jeep and secured by a bracket on each front mudguard. The jib was simple in design and easily manufactured by a field workshop. It was capable of lifting moderately heavy objects such as a tank clutch or an engine block and moving them short distances. In 1944, 1st Australian Armoured Regiment devised a similar, but more elaborate jib that involved several modifications to the jeep body and chassis and fitting a manual winch within the body of the jeep. It could reportedly lift the 559-kg weight of a Wilson gearbox out of a Matilda tank.

The 'Apparatus, Showering, Portable (Aust) No.1' in operation with the water heater in the operating position above the combustion chamber and the shower frame erected. One operator maintained the heating system while the other used the hand pump to keep a steady flow of water into the heater which could deliver a maximum of 1,135-litres of water per hour at 49 degrees C. Given the effort needed to continuously operate the hand pump, the operators swapped duties from time to time. Twelve of the jeep-mounted showering systems were issued to each mobile bath unit under command of formations equipped for tropical operations.

An expedient conversion carried out by units of 9th Australian Infantry Division during operations in North Borneo was the jeep locomotive. With all 1-metre gauge locomotives out of action and no road infrastructure, the jeep locomotives were able to maintain a steady flow of rail traffic along a network of rail lines, the distance available extending as more ground was captured from the Japanese and engineers repaired the lines, culverts, embankments and bridges. The fullest extent of the line was about 82-kms in two sections: 30-kms from Weston to South Beaufort on the south side of the Papas River, and 52-kms from Beaufort on the north side of the Papas River to Papar. (The 39-km Papar to Jesselton section was handled by restored steam locomotives.) The conversion to a jeep locomotive was simply the addition of timber baulks front and rear for pushing, additional weight inside the body to help with traction, and the substitution of the jeep's standard wheels with 724-mm diameter wheel rims scrounged from Japanese trucks. The 724-mm diameter rims were close to the diameter of the wheels on the railway rolling stock, and raised the jeep pintle into close alignment with the railway rolling stock coupling.

The wide wheel flanges allowed the jeep's track width to be reduced to match the 1-metre railway track width without having to alter the jeep's axles. The inner flanges of the wheels came close to the jeep bodywork but as the jeeps were running on rails, there was no requirement for steering. In all, some fourteen sets of wheels were manufactured. As jeep locomotives required repair or replacement, the wheels were simply transferred to a replacement jeep. The jeep's power was able to satisfactorily haul loads of around 5-tonnes at a rate of 8-kms/hour. The 'jeep trains' were successfully used for patrols, moving supplies, engineer stores and personnel, including casualties. While it was possible for a jeep to overcome the initial inertia of heavily laden railway wagons with high engine revolutions, a 'push team' of several soldiers and much slipping of the clutch, stopping took great care and the use of low gear and engine braking to initially slow the train. The jeep brakes proved useless for stopping, but could hold the train stationary after it came to a stop. Hence, the jeep locomotives quickly suffered from worn clutch plates, brake linings, and sheared-off wheel studs. The use of a jeep was not unique to the Australian Army: both the USA and the British Army in Burma utilised jeeps as locomotives. While regarded as a useful short-term solution until other means of transport or motive power become available, the Borneo jeep railway operated well into the post-war period long after operations had been returned to local administration. The jeep seen here parked at Beaufort Station was a Willys MB, chassis number MB299821. It was disposed of immediately post-war.

Willys MB, Australian registration number 147638, chassis number MB299669, being operated as a railway locomotive by 2/14th Australian Transport Platoon, 9th Australian Infantry Division, on Borneo in 1945. The jeep remained in Borneo when the railway was turned over to local administration, being officially assigned to the Allied Land Force South East Asia (ALFSEA).

Little detail is known of the jeep dump truck or 'tipper' conversion. One of the persistent problems with operations in jungle terrain was the narrowness of the tracks and how quickly the road base deteriorated with the carriage of truck traffic. The best solution for track improvement was to spread loads of gravel or crushed coral, but the in-service 4x4 tip trucks were all 2 ½ cubic yard, 3 ton vehicles which proved too large and cumbersome to get close to the forward troops. The jeep tipper appears to have been designed to dump crushed rock on access tracks too narrow and fragile for the passage of larger vehicles.

The conversion consisted of a steel tipper bin, hinged near the rear and operated by a manual winch attached between the chassis rails. The winch cable was looped through a pulley at the top of a gantry and another mounted on the front of the bin, providing a 2:1 advantage. How many jeep tipper conversions were carried out is not known. This vehicle, registration 128834, was a Ford GPW with chassis number GPW63682. It was disposed on in the immediate post-war period to FMC.

As the winch elevated the tipper bin, the gantry was hinged to lean back. The rear panel of the bin was hinged at the top, and had a latch at the bottom. Before tipping, the bottom latch was released, allowing the rear panel to move away from the bin as the angle increased. The bin was prevented from tipping right back by a chain attached to the bottom edge of the bin and the winch mounting plate.

Detail view of the winch mounting beneath the cross brace. The bin and cross brace were mounted on hardwood fillers shaped to follow the curve of the chassis rails. These were held in position by U-bolts. The winch operating handle had a ratchet to prevent the winch from reversing uncontrolled.

The bin in the raised position, showing the jeep chassis, cable pulley mounted on the front panel of the bin, and the safety chain to limit the bin's travel.

With the development of the Long Range Weapons Range northwest of Woomera in South Australia during the late 1940s and early 1950s, the jeep was called upon to undertake a number of range reconnaissance and monitoring tasks. The range was situated in an arid and semi-arid region with variable terrain including rocky plains, salt flats and high sand dunes. The standard jeep was unable to traverse the high sand dunes in particular, so modifications were made to try and improve its cross-country performance. A jeep was fitted with 9.00x13 tyres, seen in this image to the left of frame. It could climb a sand dune of 1:8 grade with minimal air pressure in the tyres which was a considerable improvement over the standard jeeps performance. Just visible in the background at the top of the dune is the 'Truck ¼ ton Special Survey', a highly modified jeep fitted with wheels and tyres from a DAP Beaufort twin engine bomber.

The 'Truck ¼ ton Special Survey' fitted with DAP Beaufort bomber wheels and tyres was developed in mid-1949. The initial conversion involved removing the bodywork, and fitting a fuel tank from an Australian Machine Gun Carrier across the rear of the chassis. The jeep's rear seat was attached to a flat floor plate which formed the new driving position. An angle steel frame supported the instrument panel. The Beaufort wheel hubs were converted to fit the jeep hubs by cutting away the wheel centre to expose the steel brake drum liner. Mild steel plates 6.35-mm thick were cut to fit the diameter of the Beaufort wheel liner and drilled to match the stud pattern of the jeep wheel hubs.

The standard jeep with the wheels dug-in due to wheel spin and the vehicle immobilised near the base of a sand dune typical of those within the Long Range Weapons Range. The 'Truck, ¼ ton Special Survey' in front had no difficulty negotiating the same sand dune. Note the wheels on the special survey jeep leave only a slight impression in the sand and retain traction on slopes as great as 1:3.

The 'Truck, ¼ ton Special Survey' was also paired with an 8-cwt trailer that had also been converted to use DAP Beaufort wheels. The combination is shown here successfully negotiating a 1.22-metre high vertical drop. The increased ground clearance of both the jeep and trailer allowed the combination to more easily traverse areas of salt bush and rock than both the standard jeep and the jeep fitted with 9.00x13 tyres.

By the end of 1949, the Special Survey jeep had been fitted with rudimentary panel work to provide protection for the engine and crew, and an enclosed stowage space. Headlamps had also been fitted. The bodywork was painted in a large black and yellow chequerboard pattern to aid in visibility. The image shows the combined jeep and trailer successfully negotiating a 1:2 side slope of loose sand.

CHAPTER 7

Experiments, Tests and Trials

Extensive experimentation was carried out in Australia, much of it in an effort to improve the jeep's performance in muddy terrain and reduce the usage of spare parts such as shock absorbers. In the majority of cases, these experiments did not lead to changes: it was difficult to economically improve on an already outstanding off-road performance. The most audacious experiment was to make the jeep fly behind a tug aircraft, and once detached from the tug, glide down and land on clearings in forward areas. Like so many of the well-intentioned experiments that consumed considerable time and resources, the 'flying jeep' or 'Fleep' did not result in a practical or viable use for the jeep.

The difficulties of traversing the muddy jungle tracks encountered in the SWPA prompted a number of inventions to improve the performance and durability in such terrain. In this instance, it was fitted with a set of rubber boots on the shock absorbers to try and reduce the instance of shock absorber failures due to the ingress of mud and water. The boots were considered suitable for service use, but were not introduced into service. Note the absence of the front bumper bar. The jeep seen here, registration 133186, was assigned to MEE at Monegeetta where it was used for a wide variety of experimental and trials work.

A special lever tool was required to attach plate retaining clips of the grousers to the wheel. A complete set of grousers could be attached to all four wheels in about 30 minutes.

Detail of the grousers including one grouser fitted with a fin, and a flotation drum fitted to a jeep wheel. Fitting a complete set of grousers and drums to four wheels took about 60 minutes. The performance of the jeep fitted with this configuration was no greater than if fitted with grousers only. The additional flotation provided by the drums was offset by the increased rolling resistance and build-up of mud.

When fitted with the full set of tractive aids, the jeep was able to complete eight traverses of a single track before becoming immobilised due to the degree of disturbance caused by the flotation drums. Conversely, the comparator jeep became bogged soon after starting the first traverse. While considerable time, effort and expense was spent on development, these traction devices were not adopted for service use.

20. GRIPS, WHEEL, FOR 6.00 x 16 TYRES

1. These Wheel Grips have been designed to replace chains in soft going where added traction is required that is not obtainable with non-skid chains. If soft mud, on soft, greasy, grassy slopes, tests have shown that wheel grips will out-perform chains when fitted to Trucks $\frac{1}{4}$-ton (U.S.A.) "Jeep," and up to a 50% greater draw-bar pull has been obtained under these conditions. Wheel grips, however, must be removed on gaining hard ground, otherwise rapid wear will result and the useful life of the grips will be reduced greatly. When used correctly, a life comparable with chains is obtainable.

2. It will be seen from Plate 35 that the wheel grip consists of 8 blade-like grips coupled by links, on each side of the tyre, which hinge on coupling pins. Plate 36 shows the Grip removed from the wheel. The grips are fitted to the tyre by wrapping around the tread in the same manner as a chain and the ends are coupled and fastened by means of a quick acting and self locking catch.

3. Wheel grips will be held in Ordnance Depots in Operational Areas for issue to such units as may be authorised by Formation Headquarters.

PLATE 35
Wheel Grips fitted to Jeep.

75

A traction aid that was adopted for service was the 'Grips, Wheel, for 6.00 x 16 Tyres', although a lack of photographic evidence suggests few were issued. The notification of their availability appeared in the January 1945 edition of the MGO Equipment Memorandum, a publication that listed new and improved items of equipment.

When the Army articulated an urgent requirement for a safe method of landing heavier equipment into forward areas, the Army Inventions Directorate initiated 'Project Skywards' with a team from the Department of Civil Aviation. Between April and December 1943, the team developed a kit based on parts from two Cierva C-30A Autogiro aircraft to enable a ¼ ton truck to be flown using a C-47 Dakota transport aircraft as the towing tug. These views of the project's 'flying jeep' or 'Fleep Rotary Wing Glider' as the project team christened it, show the Willys Model MB, registration 128276, chassis number MB197856, with the timber fairings and cloth coverings over the frame. The Fleep's driver/pilot could detach the tow rope at anytime using a lever operated from within the cabin. Once the Fleep had safely landed, the fairings and framework could be quickly and easily removed to allow the jeep to operate in the normal way. Unfortunately, the 'urgent requirement' had changed by December to 'no longer required'. As a consequence, the project was formally terminated in December 1943 just as the team were in the lead-up to flight testing. Although the Fleep never flew, all the component parts had been extensively tested on an air-worthy Cierva C-30A Autogiro, and the team leader, Mr J. L. Watkins, was confident that there was no technical impediment to the Fleep flying. The vehicle used to build the prototype was returned to Ordnance and remained in service until December 1959, when it was disposed of from Detachment 3rd BOD, Seymour.

Detail of the tubular superstructure, pylon, auto-giro rotor head and operating mechanisms for the 13-metre diameter rotor without the cloth and timber covering. The rear fuselage used on the prototype was from a Cierva C-30A Autogiro, with additional plywood pieces to widen it to the jeep's width for streamlining. Longitudinal struts were later added to support the fabric cover. The towing device mounted to the front chassis rails was based upon the standard USAAC glider towing hitch which had also been adopted by the RAAF.

The manual windscreen wiper fitted to 'Trucks ¼ ton (USA)' was considered to be a hazard in wet weather when the driver was fully occupied with steering control, gear changes and concentration on the road ahead. In 1948, the Australian Army experimented with installing a vacuum operated wiper obtained from the passenger side of a 'Truck, 3 ton, GS'. The installation included (1) the vacuum wiper motor and arm; (2) rubber connecting tubing; (3) mounting spacers inside and outside the windscreen; (4) copper tubing; (5) rubber grommet fitted into a 27/32-inch hole drilled through the dash panel adjacent to the safety strap hook eye. To allow the windscreen to be laid flat, the rubber tubing (2) was 1.09-metres long. With the windscreen upright, the excess tube was pushed through the grommet into the space under the dash panel, but when lowered, the rubber tube was eased back through the grommet to increase the length. Although considered a successful modification, and with vacuum wiper motors being readily available by removing the number required from the passenger side of 'Trucks, 3-ton, GS', the modification's introduction into service was very limited.

CHAPTER 8

RAAF and RAN Service

Relatively few ¼ ton trucks were allocated to the RAAF during the Second World War. As at November 19, 1943, the RAAF had 97 jeeps on strength, with an order pending for another 115 vehicles, making a total allocation of just 212. At that stage, this was only 75 vehicles short of fulfilling all RAAF requirements for jeeps. By March 1944, after an expansion of the number of RAAF units and a revision of their equipment structure, a further 540 'light tender' vehicles of all types were to be ordered, a proportion of which were jeeps. During November, 1944, the RAAF reported that there were 223 jeeps in service, with a further 600 on order, against a total requirement of 1,293. By the end of 1945, RAAF registrations of jeeps totalled only about 1,500. A proportion of the total RAAF jeep registrations were transferred from Army stocks during the second half of 1945, but determining the exact number of inter-service transfers is not possible from the surviving records.

RAAF jeeps tended to remain in their standard form, with few variations and adaptations. Some enterprising units converted damaged jeeps into semi-trailers. They were found so useful that the conversion became official in the post-war period, with an extensive set of conversion drawings made to facilitate modification.

The RAN received very few jeeps during the course of the Second World War. A small number were officially transferred from Army holdings in forward areas, such as to the Naval Officer in Charge on Torokina Island. It was not until the immediate post-war period however, that the RAN acquired a quantity of jeeps. These were mainly from Army stocks made surplus as a result of the cessation of hostilities. They were used at shore stations and carried aboard as a ship's 'port transport'. As ship's transport, they were small enough to easily off-load onto a dock with a ship's crane and be recovered again prior to putting to sea. RAN jeeps used for this purpose did not last very long in the salty atmosphere with most replaced by the early 1950s with the Land Rover Series 1.

Just how many ground and air crew can a jeep carry? Personnel of No.76 Squadron RAAF catch a ride along the flight line of Curtiss P40 aircraft, one aptly named 'Tropical Ulcer'. The RAAF acquired a relatively small number of jeeps through lend-lease during the war, and supplemented that number with additional vehicles transferred from Army stocks in the immediate post-war period.

Pilots of 83 Squadron RAAF squeeze into a Ford GPW, registration RAAF 212832, to move across their base at Menangle, NSW, soon after the squadron's formation in 1943. The squadron was equipped with Boomerang fighters, some of which are visible in the background, and a few Wirraway aircraft.

Warrant Officer George Baird dressed in his tropical flying kit, stands next to one of 83 Squadron's jeeps at Gove, Northern Territory, in 1944. When 83 Squadron RAAF was transferred to the Northern Territory in late 1943, Northern Territory Force allocated the unit sign serial '376'. This unit serial was originally assigned to the 704th Quartermaster Company (Aviation), USAAC, in 1942. With that unit's departure from the Northern Territory, the unit serial became available for re-issue to 83 Squadron RAAF. The black numerals were superimposed on a sky blue background. Note the unit sign is painted onto the rear left bumperette, which has been mounted upside down, and the absence of a fuel can holder on the rear panel.

A Ford GPW, registration RAAF 207199 and chassis number GPW208817, of 83 Squadron RAAF while based at Gove in the Northern Territory during the first half of 1944. The unit serial '376' has been painted onto the right side of the lower windscreen frame. In January 1944, the squadron listed an establishment allocation of ten jeeps and 3 jeep trailers, but the actual number on strength was only five jeeps. The jeep appears to be heading off on a hunting party, judging by the two single barrel shotguns visible in the vehicle. RAAF 207199 was taken on RAAF charge in October 1943, so was still a relatively new vehicle when allocated to 83 Squadron. It remained in service until December 1960.

Limits of mobility! Even the nimble jeep could get into trouble during the wet season in Northern Australia. This jeep was in service with 457 Squadron, RAAF, when it was based at Livingstone, Northern Territory. Although the RAAF did not adopt any formal unit sign system, the Army's Northern Territory Force did. From mid-1942, all units under command – Australian Army, RAN, RAAF, US Army, USAAC and civilian government organisations – were required to display unit signs on vehicles. The RAAF and USAAC unit signs were a distinctive background colour of sky blue, with the unit sign numerals in black. The heavy duty galvanised bucket is a notable addition to the vehicle's standard VOL.

A jeep serving with an RAAF unit in the SWPA. The front bumper bar has been replaced with what appears to be two pieces of light railway line with a space between.

The RAAF had no formal marking instructions other than the size and positioning of the registration number, so units applied whatever identification that seemed appropriate. This 4 Squadron jeep features the letters 'RAAF' prominently across the front bumper, '4 SQN' and '108' on the lower windscreen frame. The squadron operated mainly as an Army co-operation unit flying mostly Wirraway and Boomerang aircraft.

The jeep assigned to 'A' flight, 75 Squadron in front of one of the squadron's P40 Kittyhawk aircraft in Papua New Guinea. The 'A Flt' and 'SGT' markings across the lower windscreen frame appear to be the only identifying markings.

Preparing to pump oil and fuel into a Boomerang fighter's tanks. The three-wheeled fuel bowser trailer and two-wheeled pump unit have been towed along the flight line by one of the squadron's jeeps. Note the aerial bracket attached to the rear left of the jeep's bodywork and how the back of the canopy has been rolled up to provide better ventilation.

An RAAF jeep towing a trailer-train of several fuel bowsers at Nadzab airfield. Three other jeeps are visible in the background. Nadzab was captured in late 1943, and became a hub of intense air activity for the remainder of the war, with several RAAF and US 5th Air Force squadrons operating from its two parallel runways. The requirement to keep a constant flow of fuel up to the flight lines from the distant fuel depot required some novel means of transport!

An RAAF driver leans nonchalantly against the front left of RAAF jeep registered 207113 at Aitape, Papua New Guinea, during construction of the airfield. The jeep is a Ford GPW, chassis number GPW144059. It is still equipped with the deep wading extension on the exhaust pipe. To the right are two USAAC P-38 Lightning twin-engine fighter aircraft. In the background are RAAF airfield construction squadron and US Army personnel and equipment, including an FWD HAR-1 tipper. The tipper version was unique to the RAAF.

The jeep assigned to the Officer Commanding 'A' Flight, 4 Squadron, RAAF, negotiating a badly rutted section of the track adjacent to the airstrip on Labuan Island in October 1945. By this time, the squadron was operating CAC Wirraway aircraft for general reconnaissance, transport and leaflet-dropping.

Flight Lieutenant E.V. (Val) Walliker, Commanding Officer, 4 Squadron RAAF sitting in his Ford GPW, registration RAAF 207083, on Labuan Island in October 1945.

An RAAF jeep stands by as an RAAF Consolidated PBY5A Catalina flying boat is incinerated on Moratai Island in June 1945. This may be A24-91 (ex-USN46488) that according to an eyewitness account, accidentally caught fire while parts of the aircraft were being washed down with aviation fuel – a common practice to remove oily grime from the matt paintwork. In the background is a B-24 Liberator bomber.

Sergeant Frank Kuffer, a Medical Orderly serving on Bougainville Island with 10th Local Air Supply Unit standing next to the unit's Ford GPW registration RAAF 213577, chassis number GPW151788. In addition to the regulation 'RAAF 213577' the unit has also painted the red cross symbol on the left and right ends of the front bumper, on each side below the door opening, and on the rear bumperettes.

Supplementing the rations with some of the local wild fruits. Frank and an associate using a jeep to scrounge bananas and mangoes from the jungle on Bougainville Island. Note the damaged front bumper bar.

Another variation on marking an RAAF jeep. This one has two Australian Pacific theatre roundels (blue outer, white centre) painted on the lower windscreen frame, and a large red cross symbol on the bonnet. The registration is also on the flat of the bonnet rather than on the rounded edge or sides.

Two jeeps parked outside the outpatient clinic and ward buildings of the RAAF Medical Section on Bougainville Island. The near jeep is standard while just visible behind it is an ambulance jeep equipped with the three-stretcher, two-tier stretcher supporting gear. Note the large red cross symbol on the rear right.

Frank Kuffer's jeep even featured on his mail home! Interestingly he has featured the jeep in disruptive camouflage.

Two happy RAAF personnel sitting in a Ford GPW, chassis number GPW229830, outside the squadron crew hut. The jeep is one of a number taken on charge by the RAAF in late 1945 after the cessation of hostilities. It remained in service until early 1954. Note the size and position of the 'RAAF' and registration numerals, and the absence of the canopy bows. The vehicle is finished in the wartime Khaki Green No.3 lustreless finish.

In the late evening on 5 May 1945, RAAF 212819, a Ford GPW chassis number GPW192819 on issue to 84 Squadron and driven by Sergeant Reg Peterson, was returning to camp at Ross River, Queensland, from Townsville, a distance of about 40 kilometres. While travelling at high speed, the jeep struck a rock which caused it to leave the road and flip. Peterson, an experienced 'Driver, Motor Transport', was killed instantly while a passenger, Sergeant Watts, suffered bruises and abrasions. Note the position of the Tare weight information along the edge of the fuel tank well. The cause of the accident was officially listed as excessive speed.

An RAAF 'Truck, ¼ ton' on operational service again, this time in Korea with 91 (Composite) Wing, elements of which were stationed at the airfield at Kimpo. It is a Ford Model GPW, chassis number GPW264455, registration number RAAF 215812 which was taken on charge at the end of Second World War on 20 September 1945 – one of a number of ¼ ton trucks received by the RAAF at that time. It was disposed of in October 1954. Note the large five-pointed star on the bonnet. The US star was adopted for recognition purposes by all UN forces in Korea. In the foreground is a new General Motors-Holden's Ltd utility – one of the new breed of general service light cargo vehicles purchased by both the RAAF and the Australian Army in the post-war period. This vehicle, RAAF 216902, was received into service on 19 September 1952, and was finally disposed of in August 1962. In the background is a Gloster Meteor jet aircraft operated by 77 Squadron RAAF.

Pilot Officer Randle Green and Flying Officer Dick Turner playing a game of draughts on the bonnet of the 77 Squadron Commanding Officer's personal jeep on the flight line at Kimpo, Korea, in 1952. The jeep is finished in the RAAF post-war gloss blue-grey paint scheme. The additional insignia across the lower windscreen frame leaves no doubt that this is Squadron Leader Dick Cresswell's personal transport. Note the additional height piece with the rounded end along the top edge of the windscreen frame. This simple addition raised the canopy a few more centimetres above the driver's and front passenger's heads making travel with the canopy erected more comfortable.

An RAAF jeep at a refuelling and maintenance depot in Korea. Note the addition of light-coloured canvas side curtains and doors to make travel in bad weather more tolerable. A CMP Cab 13 water tanker truck can be seen nearby, with other vehicles including a jeep and staff cars in the background.

There are several things to note about this Ford GPW, registration RAAF 216116, chassis number GPW230009. It has the later rotary light switch on the dash and is missing the rear vision mirror arm. The canopy has a reinforcing strip stitched to the outside above the centre canopy bow, and the front of the canopy has been raised about 75-mm by a lateral piece of timber held in position by the addition of two short sections of steel strap welded to the top edge of the windscreen frame. The driver's seat appears to have a padded leather seat cushion with lift-dot fasteners around the edge. The bonnet left side has four screw heads visible, indicating the inside of the bonnet is fitted with the grease gun bracket. The windscreen lower frame appears to be fitted with the rifle bucket mounting brackets, one of which is being used as a convenient place to bulldog clip some papers.

The enterprising personnel of 25th Air Stores Park 'found' a damaged and abandoned jeep during the unit's time on Noemfoor Island in 1944, and converted it to a semi-trailer. The trailer was made using the rear section of another jeep chassis from just forward of the rear springs' front spring pivot bracket. Two longitudinal timber supports were shaped to fit the curve and held in position by U-bolts. A timber deck was attached to the top and a simple fifth wheel at the front. The rear body of the jeep was cut off and the lower half of the fifth wheel attached to the intermediate rear frame cross-member and the rear frame reinforcement. The conversion was considered to be a very valuable addition to the unit's transport and moved with the unit to Labuan Island in June 1945, where it is seen here being used for moving crates of aircraft spares. When 25th Air Stores Park was disbanded in November, remaining stores and equipment were transferred to 30th Air Stores Park which disposed of surplus stores and vehicles in late December by dumping at sea. As the jeep tractor and semi-trailer were excess to inventory and not a registered RAAF vehicle, this was its likely fate.

An RAAF jeep semi-trailer being used as a hearse. Attempts were made by RAAF personnel at unit level to convert a jeep into a semi-trailer version, including this one. RAAF 212726 was a Ford GPW, chassis number GPW192723, which was taken on charge on 15 November 1944 and remained in service until June 1955. This conversion involved removing the bodywork from behind the front seats and inserting a straight panel. A turntable plate was mounted on lateral supports fixed to the jeep's chassis between the centre frame cross-member and the intermediate rear cross-member. The semi-trailer was manufactured as a flat platform supported by two longitudinal members, while the rear springs, axle and wheels were all repurposed jeep trailer parts. The overall post-war paint finish was RAAF blue-grey synthetic gloss enamel. Realising the utility of the semi-trailer configuration in the immediate post-war period the RAAF designed a much more elaborate goose-necked semi-trailer conversion which utilised readily available commercial parts. It had an overall length of 4.3-metres, with a load-bed length of only 2.6-metres, the remaining 1.7 metres being a narrow tongue forward of the gooseneck which included the turntable. The load bed was not flat, but interrupted by wheel wells on each side. Although RAAF Technical Services finalised the specifications and drawings, it appears none were built.

The RAN did not officially receive any jeeps during the Second World War, although a few were transferred locally from Army stocks in the Pacific Islands for the use of specific personnel, such as RAN liaison officers. Once received, the navy evidently made sure everyone knew who it belonged to! This prominently marked vehicle, probably finished in light grey overall, is transporting three naval personnel on Bougainville Island.

Two jeeps and a 12-cwt utility lashed to the deck of HMAS Manoora. Like the RAAF, the RAN did not adopt any formal system of unit markings. Manoora's jeeps were simply marked 'RAN' centrally on the lower windscreen frame, and a small 'M' followed by a numeral. The 'M' number fell in the same sequence as the ship's landing craft, which were numbered from M1 onwards.

The immediate post-war period saw a small number transferred to the RAN for use in shore stations and especially as ship's transport. This RAN jeep has been repainted in an overall grey paint finish with what appear to be white-wall tyres. It has 'RAN' painted in white on the right bumperette, and the registration number across the rear frame crossmember below the pintle hook. The canopy should have been folded back or removed prior to lifting, as it is being damaged by the sling pulling the sides in.
(See also Appendix 6)

CHAPTER 9

Trailers

When the jeep first entered service with the Australian Army in mid-1942, utilising it for towing another vehicle or trailer was not initially authorised. In late November 1942, the Army Commander-in-Chief issued General Routine Order O-583 which stated that " *(a) approval is given for Trucks, ¼ ton 4x4 (jeep) to be used as a towing vehicle, within its capacity. (b) The trailer or towed vehicle shall not be loaded with a weight exceeding 8 cwt and the total weight of the towed vehicle and load shall not exceed 12 cwt even on good made roads. (c) When used for cross country work the above weights shall be reduced to ensure that no overload or severe stressing takes place.*"

Even before this approval was given, the Australian Army was planning to introduce a trailer primarily designed to be towed by a jeep. In accordance with the government policy to manufacture as much equipment as possible in Australia, the first Australian 8-cwt capacity trailer was a local design drawn up by the Army Design Directorate (ADD). This design, the 'Trailer, 2-wheeled, 8-cwt (Aust) No.1', while simple, lacked the robustness required for service use. It was followed by two improved versions, the No.2 and No.2A trailers. These also lacked sufficient strength for rough use, so the decision was made to supplement trailer holdings by importing US-manufactured ¼ ton trailers under the lend-lease scheme while simultaneously manufacturing a trailer of similar design in Australia. Bids for allocations of US ¼ ton trailers were first placed in late 1943 which began arriving in Australia in early 1944. Bids totalled 4,600 US trailers. By the end of the war, US authorities had released 4,461 trailers to Australia. Although the US Army classified the jeep cargo trailer as a ¼ US ton capacity, that is, 500-lb or 227-kg, the Australian Army classified the carrying capacity as 8-cwt, that is, 896-lb or 406-kg – the same as the Australian-manufactured trailers

The first of the Australian-manufactured trailers based on the US trailer design was the 'Trailer, 2-wheeled, 8-cwt (Aust) No.3' which was quickly followed by an improved version, the No.3A. These began delivery from contractors in late 1943. The last locally-designed trailers based on the US trailer was the 'Trailer, 2-wheeled, 8-cwt (Aust) No.4'. This was a lighter-weight construction than the No.3A. Deliveries from the two manufacturing contractors began in June 1945.

The RAAF obtained a small number of trailers based on the Army's (Aust) No.2A trailer, but without the cutouts in the end panels for stretcher handles or the drawbar parking support. These were manufactured by F.Morse Pty Ltd in Warrnambool, Victoria. The RAAF returned to the same company in May 1944 to have another trailer manufactured to RAAF specifications. This was the 'Trailer, ¼ ton, Morse, Model 3' which was delivered in batches between August 1944 and May 1945.

The Australian jeep trailer with the 'covered wagon' type canopy erected.

Initially, the Australian Army did not order US pattern ¼ ton trailers for towing by jeeps, preferring to have a locally designed trailer manufactured in Australia. The 'Trailer, 8-cwt, (Aust) No.1' which was originally simply designated as 'jeep trailer'. It was introduced into service at the end of 1942 and was a simple design of a fabricated steel angle and flat strap frame with sheet steel sides and ends. The sheet steel floor was corrugated for additional strength. The front end panel was fixed while the rear panel was removable for loading. Both ends had apertures to fit the handles of a standard Army stretcher. A removable canopy frame of four half-round individual hoops was fitted and covered with a canvas canopy which could be laced securely to the body. The draw bar was tubular with flat angle braces, and the towing eye was a 'U' of 12.7-mm round bar welded to the end of the draw bar.

Pontiac motor car springs and shackles were used, together with a tubular axle of 50.8-mm diameter steam pipe with Bantam stub axles welded into the ends. The axle was mounted to the springs on timber perches. It was equipped with Bantam car pressed steel wheels with 5.00 x 16 conventional pattern tyres with a spare wheel assembly carried under the trailer body forward of the axle. The advantage of the design was that the wheels, axle, springs, drawbar and canopy were all readily detachable and could be stowed inside the body to minimise the freight volume.

The underside of the trailer showing the spare wheel mounted beneath the body forward of the axle. Testing of a fully laden trailer indicated that the maximum safe load was 995-lbs, or 8.9-cwt (451-kgs) which resulted in the trailer being classed as an 8-cwt capacity trailer.

The prototype Australian jeep trailer behind a jeep, showing that the trailer sat level when both the jeep and trailer were unladen. Note the spare wheel is not yet mounted.

The prototype with the four hood bows in position.

An Aust No.1 trailer in use in Papua New Guinea. Operational experience during 1943 indicated that the number of trailers should be increased substantially to one for every two jeeps on issue.

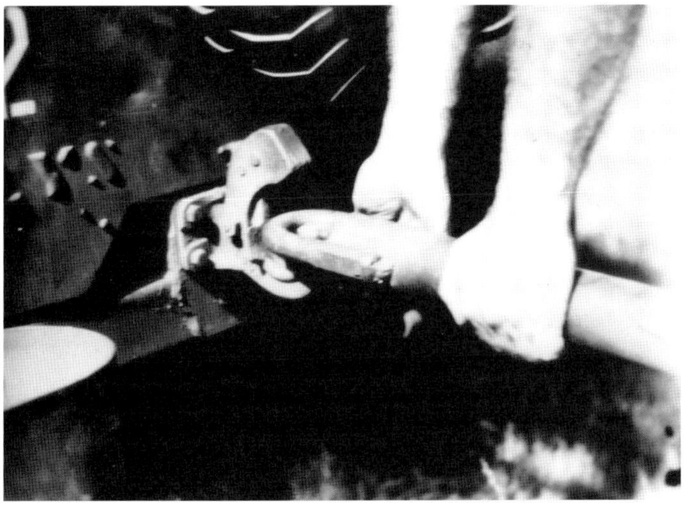

One of the faults with the first design of Australian jeep trailer was the excessive size of the towing eye. The eye allowed far too much movement within the pintle hook, which resulted in a persistent and loud rattling particularly on rough surfaces, and when braking or changing gears. Only about 90 of the No.1 jeep trailers were built by the NSW Line of Communication Area Workshop and Peter Brothers before the design was improved. In any event, operational use soon determined that the timber perches in particular were not sufficiently robust, so a complete kit of suspension, axle and wheel assemblies was issued in May 1944 to convert the No.1 trailer to the (Aust) No.2A configuration. (See Appendix 7-2)

The improved design was designated the 'Trailer, 2-wheel, 8-cwt, (Aust) No.2'. The body was identical with the Aust No.1 while the wheels, axle and springs were all upgraded. The springs were more robust eight-leaf springs built specifically for Army trailers, and the axle was a 35-mm square section bar fitted with Australian manufactured wheel hubs, pressed steel wheels and 6.00 x 16 conventional pattern tyres. About 500 were manufactured by Peter Brothers. In June 1943, Australian trailers were being delivered from the contractor at a 'satisfactory rate'.

Further improvements were made in a second series of Aust No.2 trailers, sometimes designated as 'Trailer, 2-wheel, 8-cwt Aust No.2A'. The 2A version had a spring-loaded drawbar and the round towing eye was fitted with a removable take-up piece. When the take-up piece was installed, the towing eye was a snug fit into the jeep pintle hook, but when removed, the trailer could be towed behind a vehicle with the larger War Department pattern pintle hook. An adjustable height parking support was fitted to the 2A drawbar which could be folded up under the body when not in use. The version shown here was manufactured for the RAAF by F.Morse Pty Ltd in Warrnambool, Victoria, and differed from its Army equivalent by not having the parking support or the apertures for stretcher handles in the front and back panels. Like the earlier versions, the (Aust) 2A could be dismantled and packed within the trailer body, as can be seen here packed into an open timber crate ready for delivery to an RAAF store. About 50 of the RAAF version of the (Aust) 2A trailer were manufactured. Over 20 are seen here packed aboard an RAAF Maple Leaf Series 16, 7-ton 'flat top' semi-trailer, registered number RAAF 200216, chassis number 4116M1022.

Come and get it! Ambulatory patients at an RAAF Medical Clearing Station in New Guinea crowd around an Aust No.2 trailer while dishing out their lunch. The 'lunch cart' filled with hot boxes and a tea bucket was moved from ward to ward each day.

A No.2 trailer modified into a mobile saw bench by 12th Field Company, Australian Engineers, being prepared for movement to a new location near Lae, New Guinea in October 1944. The trailer's left side panel has been removed and the saw bench attached to the forward section of the trailer. The un-protected saw blade appears to be powered by a direct drive from a pulley attached to the fly wheel of a jeep engine, though how this is aligned in operation is not clear. It is possible that the engine pack is slid outside the left side of the trailer in order to align the pulley with the saw blade drive, and stowed within the trailer when ready for movement. Note the support bracket for the starter motor shaft bearing which would normally be within the jeep's bell housing. The throttle linkage has also been modified. The rear of the trailer is kept level by being propped on pieces of timber on Australian 4-gallon fuel drums. The original wheels have been replaced with jeep two-piece combat rims.

A heavily laden jeep and 'Trailer, 2-wheel, 8-cwt (Aust) No.2' being ferried across a river aboard a raft made from 44-gallon fuel drums lashed together with a timber and arc-mesh floor.

A jeep trailer-train of US pattern trailers with a single Aust No.2 trailer bringing up the rear. A report in November 1943 noted that Australian pattern jeep trailers were '... quite unsuitable. Suffered from excessive spring breakages, bending of perch, distortion and finally virtual destruction of body due to poor construction. Springs and wheels are not interchangeable with similar jeep parts' and by February 1944, the No.2 trailers in use in Papua New Guinea were reported to be in a '... state of collapse. Bodies are bashed out of shape, chassis bent, axles fatigued and consistent repair to the tow eye has rendered it beyond further capability of repair. ... The American trailer has been used to a small extent by our forces and has proved reasonably satisfactory.' The trailer train required the addition of a towing pintle to the rear of a trailer, but this could not be added to the Aust No.2 due to its construction and lack of structural strength. Trailer trains such as this one being towed by a tracked tractor were found to be useful when roadways became impassable due to mud, but only for a short time as the tracked tractor invariably caused even more damage to the road base.

The superiority of the US ¼ ton trailer design had been recognised as early as the middle of 1943, and the decision made to base all future local production of Australian 8-cwt trailers on that design. The US design incorporated rear and side reflectors, a spring-loaded towing eye, a fold-up stand for parking, a park brake and rear brake, stop and blackout lights which were powered through a plug socket located on the left of the jeep's rear panel. Bids were placed with the Australian War Supplies Procurement (AWSP) organisation in the USA to procure quantities of the US trailer under the Lend-Lease agreement in order to supplement local production. At least 4,166 US pattern ¼ ton trailers were subsequently supplied to Australia. These were designated as 'Trailers, 2-wheeled, 8-cwt (USA)'. While the authorised carrying capacity in US service was ¼ ton, that is, 500-lb or 227-kgs, the Australian Army, following load testing at MEE Monegeetta, authorised 8-cwt, that is 896-lb or 406-kgs – a considerable increase.

The underside of the Mobile Office trailer, showing the method of stowing the benches and support poles. The support brackets along the sides and rear of the body were sealed to prevent the ingress of water thereby preserving the ability of the trailer to be floated across streams.

The trailer with all benches secured to the mounting brackets and the canopy support frame erected. The conversion included a large canopy which was placed over the support frame and secured to provide an awning over the benches and trailer. Office equipment – including chairs, boxes of forms, writing materials, and so on – were stowed within the trailer and were easily accessible once the trailer was fully deployed. Note the cut-outs in the side benches to accommodate the curved mudguards.

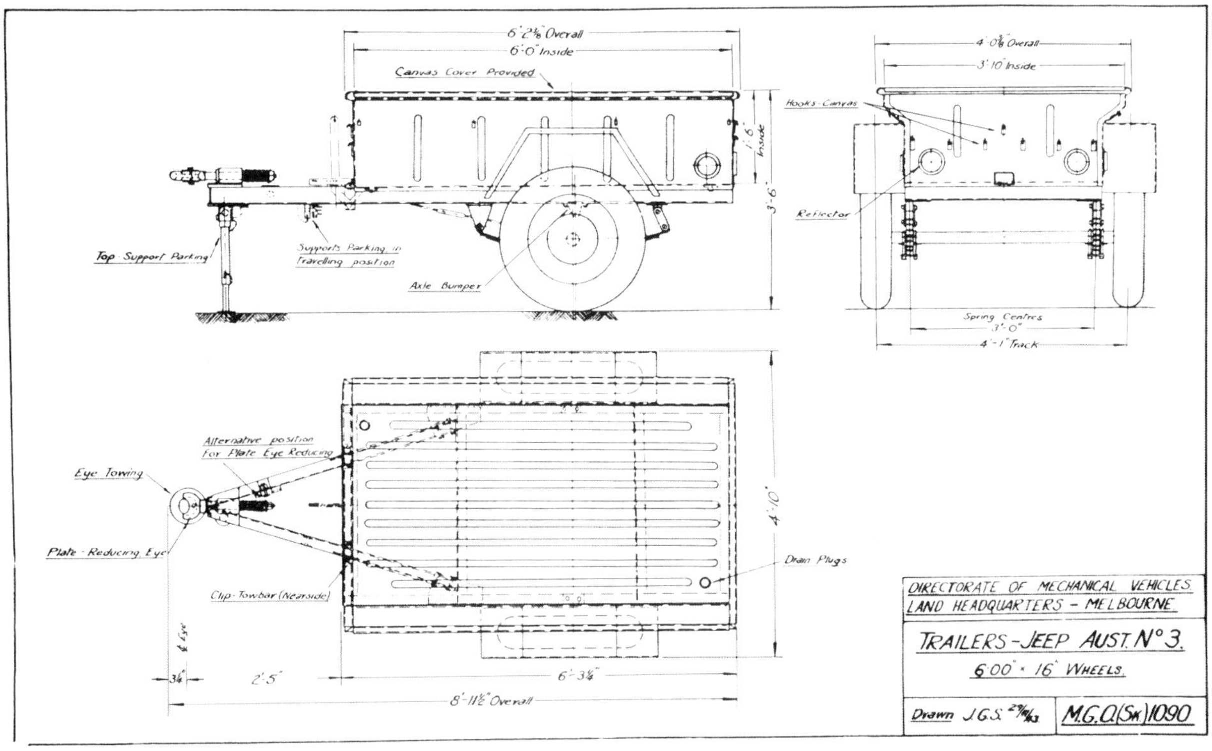

The first Australian 8-cwt trailer based on the US design was the 'Trailer, 2-wheeled, 8-cwt (Aust) No.3'. Design work was completed by August 1943. The early design included a handbrake that could be operated from within the towing vehicle and brackets on the trailer body to secure the legs of one or two stretchers. By November 1943, a pilot model had been subjected to trials including a flotation test in Corio Bay on a calm day that indicated the No.3 trailer loaded with 8-cwt had over 200-mm of freeboard. The trailer also handled a choppy sea with a load of 4-cwt without being swamped. Deliveries of Aust No.3 trailers began toward the end of November 1943, but for some unknown reason, deliveries to units overseas were slow. The first batch of 64 trailers dispatched overseas arrived at Milne Bay, Papua New Guinea, in early February 1943.

The similarities in the design of the Aust No.3 trailer with the US ¼ ton trailer are evident. The body was of welded sheet metal with an integral chassis. The chassis was made of rolled steel sections welded together and to the underside of the body tub. The top edges of the body tub were rolled and welded to make a tubular top edge which provided rigidity and minimized the risk of denting when heavy loads were being loaded. The trailer was waterproof and could float, which also meant liquids would be unable to drain from inside. To overcome this, drain plugs were incorporated in diagonally opposite corners of the floor. The body panels were pressed with vertical ribs to provide rigidity. An angular sheet steel mudguard was fitted over each wheel and was joined to the body panel at the bottom edge of the chine. Both the shape and point of attachment were salient features of the Australian No.3 trailer.

The problem was exacerbated when the jeep was laden with 363-kgs and the trailer was unladen which resulted in the trailer having a distinctly nose-down attitude. This made backing quite difficult.

Testing showed the problem could be overcome by mounting the towing eye 114-mm lower. With four men in the jeep and the trailer laden with 406-kgs (8-cwt), the lowering of the towing eye resulted in the trailer tracking more or less level.

The Australian No.3 and 3A trailers had a cast nomenclature plate fixed to the back panel by two rivets. While the information displayed on the plates was the same for the four manufacturers, there were minor differences in layout. This Aust No.3 plate was the type used by Charles Hope which left a blank raised block to be stamped with the trailer model and serial number. This is serial CH62, which equates to registration 148603. The trailer was disposed of to the Post Master General's department after the Second World War. The Aust No.3A nomenclature plate used by FMC had no raised blocks with the serial number simply stamped on the vacant space. Serial 342 was registration 133128. In the Army's registration ledger, the serial number was recorded as 'JT342'. The official use of the word 'jeep' on the nomenclature plate is of note. By 15 July 1944, 801 Aust No.3 and 3A trailers had been delivered from the contractors of a final production total of 1,709.

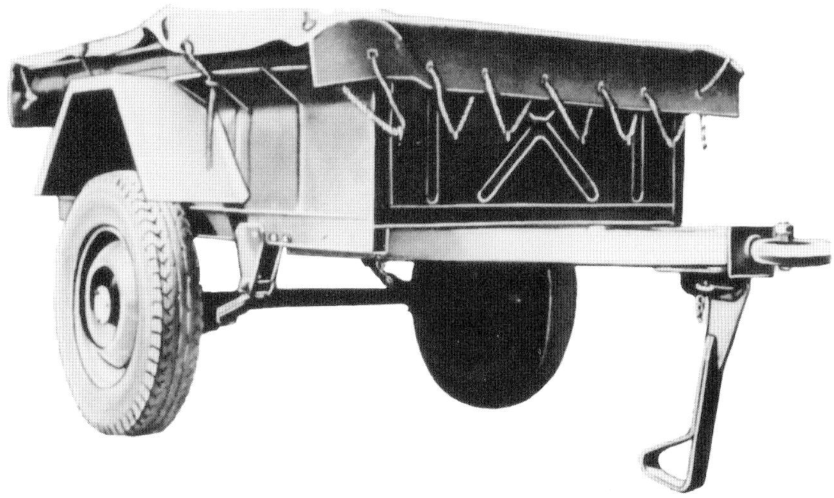

As a result of an overall shortage of 8-cwt trailers in mid-1944 and with no prospect of further allocations from America until July 1945, Army was forced to revert to local manufacture. An improved version of the Aust No.3A was designed under the existing V124 project number. The new design took account of the experience with all the existing models of 8-cwt 2-wheeled trailers. The result was the 'Trailer, 2-wheeled, 8-cwt (Aust) No.4'. Preliminary discussions for the manufacture of the new design took place with the Department of Munitions in October 1944 followed by an order in late November for 3,300 trailers. Munitions awarded manufacturing contracts in January 1945 to GMH and FMC, with each to manufacture 1,650 No.4 trailers. Deliveries were expected to commence in April 1945 at a rate of 200 trailers per week. The Aust No.4 trailer was similar to the Aust No.3A but was 11% lighter and simpler to manufacture. It used a locally manufactured tubular axle and wheels. No spare wheel was carried as the wheels were interchangeable with the jeep towing vehicle, hence the spare from the jeep could therefore be used on the trailer in an emergency. The spring loaded towing eye was mounted lower so the trailer was close to level when both the jeep and trailer were fully loaded. A drain hole with a screw plug was located low in the rear panel. A parking stand was mounted under the front of the drawbar. This folded back level with the drawbar when not in use. The towing eye was sized to allow towing by vehicles with the larger War Office type towing pintle or the smaller jeep pintle. A demountable take-up piece fitted inside the eye to reduce the size for the jeep towing pintle. The Aust No.4 was characterised by the 'X' shaped strengthening ribs pressed into the front and back panels, and the angular mudguards mounted at the top of the angled section of the side panels. The Aust No.4 also included a single tubular canopy hoop that mounted into sockets attached at the midpoint of the top edge of the body on the inside. The canopy hoop helped to stiffen the body and was arched so that water would shed off the canvas canopy. Deliveries of Aust No.4 trailers were held up due to delays in the delivery of axles, hubs and bearings. FMC completed their pilot model registration 165503 on 21 April 1945, while GMH delivered their pilot model registered 152691 on 28 May 1945. Approval for manufacture was provided almost immediately and without the usual trials phase. Deliveries of trailers from both manufacturers commenced about 9 June 1945. The rate of delivery was, however, governed by the slow arrival of wheel bearings from the USA. GMH deliveries were also delayed due to the discovery of a small number of defective shackle bolts.

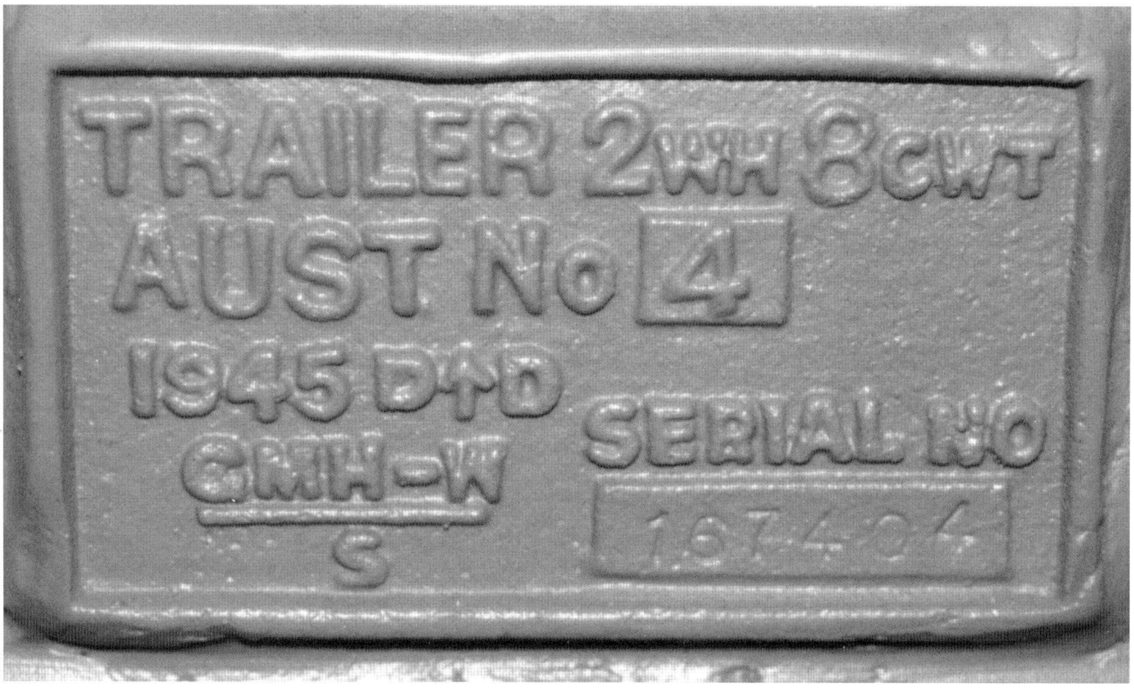

A similar cast plate to that used on the Aust No.3 and 3A trailer was attached to the back panel of the Aust No.4 trailer. The word 'jeep' was not used, with the full Army nomenclature in an abbreviated form used instead. This plate was the design used by GMH. The symbol 'GMH-W' indicates the GMH plant at Woodville while the 'S' indicates the plant is in the state of South Australia. The serial numbers of GMH manufactured Aust No.4 trailers were the vehicle registration number.

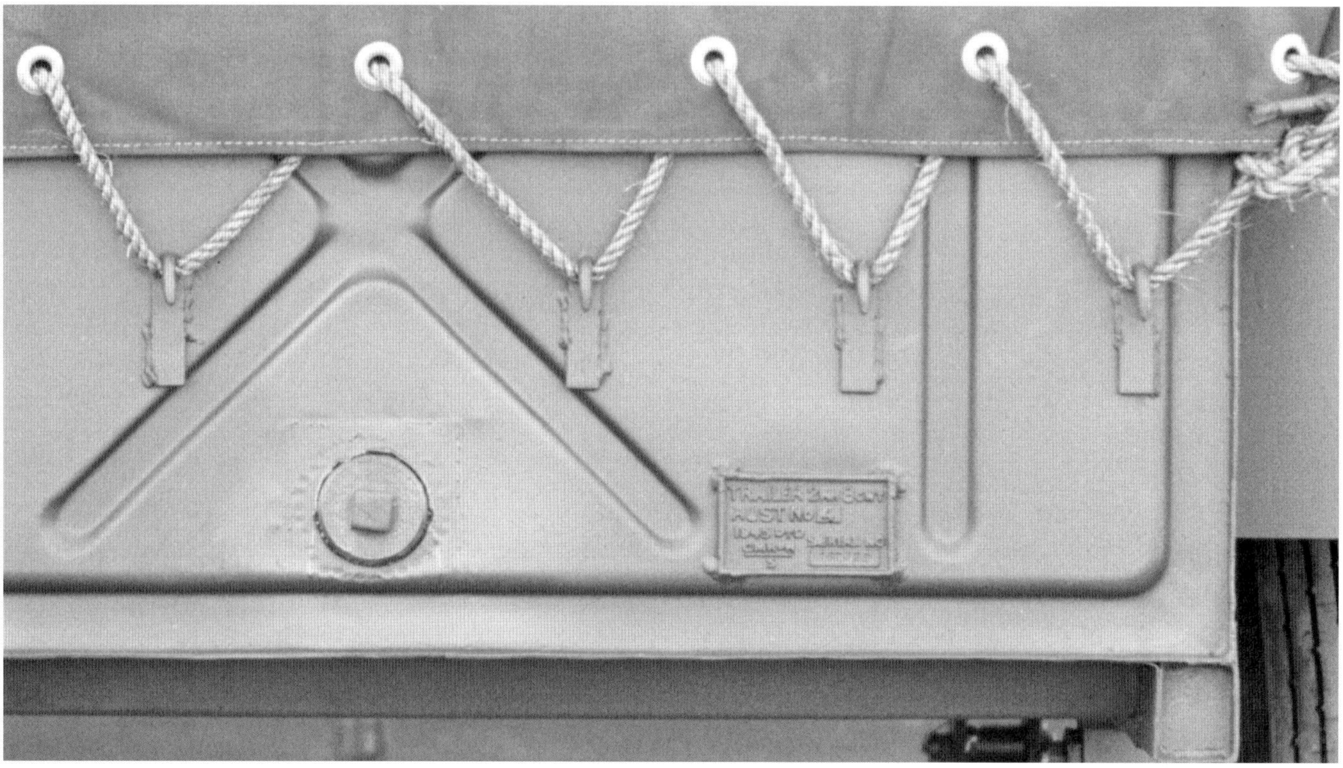
Details of the rear panel of a No.4 trailer manufactured by GMH showing the placement of the nomenclature plate to the right of the drain plug, the 'X' shaped pressing to provide strength, the type and spacing of the tie-down hooks and the depth of the overhang of the canvas canopy. This is registration 167329 which was disposed of to Dominion Motors soon after the end of the war.

No.4 trailers awaiting delivery from FMC in Geelong, Victoria. Note the placement of the nomenclature plate on the lower right of the back panel, and the drain plug in the lower centre. The registration numbers were assigned in blocks to the two manufacturers and stencilled onto the back panel as part of the manufacturing contract. Paint was satin finish Khaki Green No.3 overall. The canvas canopy and canopy bow were supplied by the manufacturer. FMC assigned a sequential number to the No.4 trailers. Registration number 166523 was JT1468, while registration 166524 was JT1440. Both were disposed of to Stokoe Motors soon after the end of the war.

'Trailers, 2-wheeled, 8-cwt (Aust) No.4' awaiting delivery from the holding yard at GMH Woodville, South Australia. Two lines of No.9 artillery tractor bodies can be seen in the background.

An Aust No.4 trailer slung from the jib of a Diamond-T 4-ton medium wrecker. The drawbar has been removed and is balanced against the suspended trailer. Note the way the floor became stretched, dented and scratched with use, the distortion showing up the underside lateral supporting ribs. One of the complaints about the Aust No.4 trailer was its lack of robustness due to the light-weight construction. The Assistant Superintendent of Design admitted that in designing for lightness some reduction in robustness had been made, but that the trailers' in-service use indicated the working life appeared to be equal to the 'Trucks ¼ ton (USA)' which towed them.

Despite earlier indications that further allocations of US 8-cwt trailers would not be made before July 1945, trailers both new and reconditioned began arriving again in May 1945.

An overloaded US pattern ¼-ton trailer being towed by an equally overloaded jeep! Soldiers enjoying a 'third class ride' rather than a 'first class walk'!

Experiments were conducted at MEE with trailer trains using a 'Tractor, Tracked' for haulage. Here the offset of the rear trailer to the tractor is 6-feet (1.83-metres) on a moderate side slope. Fitting a pintle hook to the rear coaming of jeep trailers to allow 'trailer trains' was not approved for service. These are Australian No.3 trailers. Registration 140899 was manufactured by GMH with the identification number JT656. It was disposed of soon after the end of the war to Carpenter and Company in New Guinea.

While not officially approved, 'operational expediency' saw troops adapt trailers into trains towed by a tracked tractor, in this case, a bulldozer. These trailers are a mix of Australian No.4 and US pattern trailers and are carrying 44-gallon capacity fuel drums.

The harsh, muddy conditions in forward areas often necessitated some extreme measures. These trailers carrying ammunition boxes, have all had their mudguards removed to prevent the build-up of mud between the inside of the guard and the tyre. They are being towed in a trailer train by a bulldozer.

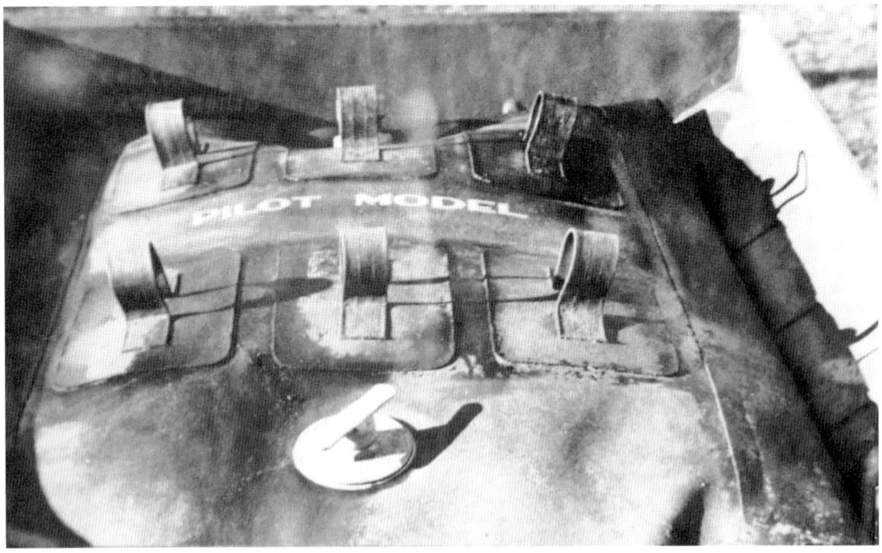

Transporting potable fresh water to troops in jungle areas accessible by rough and narrow tracks was not possible using the standard 3-ton water truck, so various methods were tried using jeeps and trailers. A collapsible rubber lined canvas water bladder with a tare weight of 51-kgs that fitted within an Australian No.3 or 3A trailer was developed in mid-1944. While the total volume was 608-litres, it had a nominal capacity of only 363-litres and an actual capacity of 354-litres, the quantity of water being dictated by the total load weight in the trailer. Trials with the pilot model over a distance of about 6,500-kms showed that the construction was sufficiently robust for service use but towing was seriously affected by the sloshing of the water within the bladder. This severely limited the safe towing speed particularly when cornering or braking. The water was also tainted by the rubber, but this lessened with each refill. Further development was subsequently abandoned in favour of a trailer-mounted 80-gallon water tank.

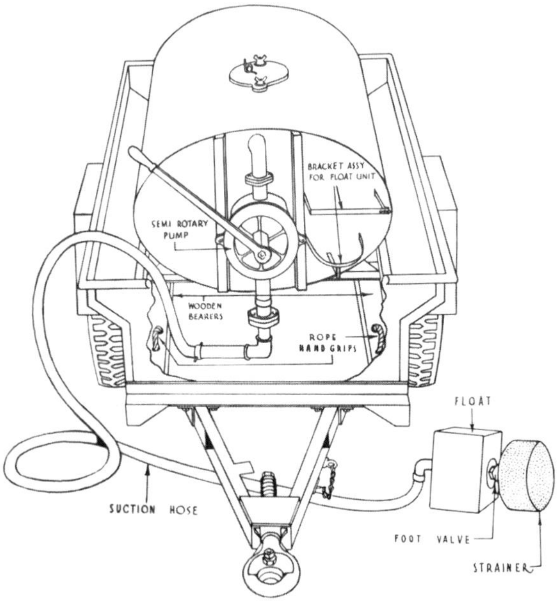

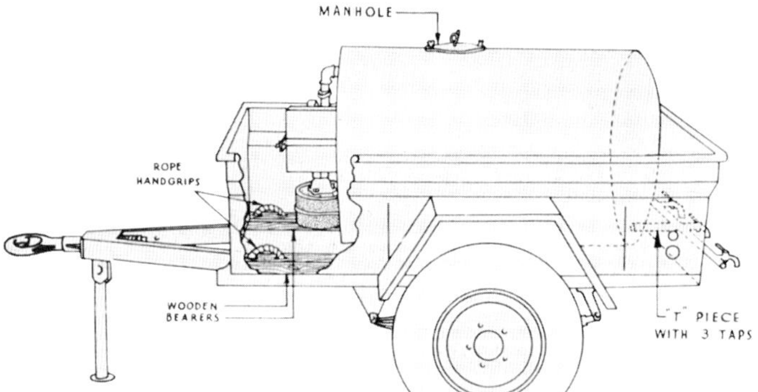

The method adopted was the 'Kits, Tank, 80-gallon, Water' which was designed to fit within the Australian No.3, 3A or 4 trailers and the US type 8-cwt trailer. The kit was manufactured by FMC. It consisted of an eliptical tank manufactured from 14 gauge (1.897-mm) mild steel with welded seems. A circular hole with a cap provided limited access to the interior of the tank for maintenance and as an alternate filling point. A pair of longitudinal wooden beams the length of the inside of the trailer supported the tank and prevented movement. The beams were fitted with rope handles for lifting. An Epex No.5 semi-rotary manually operated pump was mounted on the front of the tank. The pump was connected to a 6-metre, 19-mm diameter flexible suction hose with a float, foot valve and inlet strainer. This could be placed into any convenient supply of water such as a lake or stream, and the tank filled using the hand pump. At the rear of the tank was the water outlet just above the bottom of the rear panel. The outlet was a 'T' piece with three taps which mounted on the outside of the trailer's rear panel. To install the outlet 'T' piece, a hole had to be cut in the trailer rear panel of the Aust No.3, 3A and the US type trailers, but not the Aust No.4 where the drain hole was used. The rear of the tank also had a stowage box for the accessories such as the water test kit, taste removal tablets, spare washers and hose clips, and a tank cleaning brush. Issue of 'Kits, Tank, 80-Gallon, Water' began in October 1944.

Private McCormick of 2/3rd Australian General Transport Platoon removes the tank cap in preparation for gravity-feeding water from an elevated holding tank at Cape Karawok, Papua New Guinea in 1945. The elevated tank was filled by a siphon from a stream further up the slope. The trailer, registration 74842, is an Australian No.3 manufactured by GMH with frame number JT-254. Instead of the 'T' piece and three taps, the tank outlet is fitted with a short length of hose and a control valve. The markings are 2-3 over 91 on a diagonally divided background colour of red in the upper left and green in the lower right. The formation sign is the kangaroo over a boomerang in white on a black background of the 6th Australian Infantry Division.

Filling a trailer-mounted tank with sea water was not the use the Army had envisaged for the water tank kit! Troops of 11th Australian Infantry Battalion at Malaguna, New Britain, in September 1945, were using the water to dampen the roads in the area in an effort to reduce the dust. Note that the kit has been installed into a US pattern trailer. In the background is an ALC40 landing craft. These 19.5-metre long sea-going transport vessels had a capacity of 40-tons of stores, personnel or vehicles. They were built by FMC at Geelong, Victoria and Brisbane, Queensland.

In May 1944, the RAAF placed an order with F.Morse Pty Ltd of Warrnambool, Victoria, for a trailer with a cargo bed of 2.03-metres x 1.14-metres wide. A pilot model was manufactured, tested and various improvements incorporated in the manufacture of a further 142 trailers. The trailers were delivered in batches between August 1944 and May 1945. They were assigned registrations RAAF 211072 to 211214. Construction was welded sheet steel with all body seams waterproofed. Two drain holes with plugs were located in diagonally opposite corners of the body. Load capacity was rated at 254-kgs (560-lb). The RAAF nomenclature was 'Trailer, ¼ ton, Morse, Model 3'.

During the post-war period, the RAAF converted a number of Australian No.4 trailers to magnetic runway sweepers with an electro-magnetic coil suspended beneath the axle and just above tarmac level. A generator mounted inside the trailer provided the power to the coil. The trailer was towed at a slow, steady speed along the airfield runways to catch any metal objects that might puncture aircraft tyres or be thrown up and damage airframes and engines. Runway clearance became even more important with the introduction of the first jet aircraft, the De Havilland Vampire, in 1949.

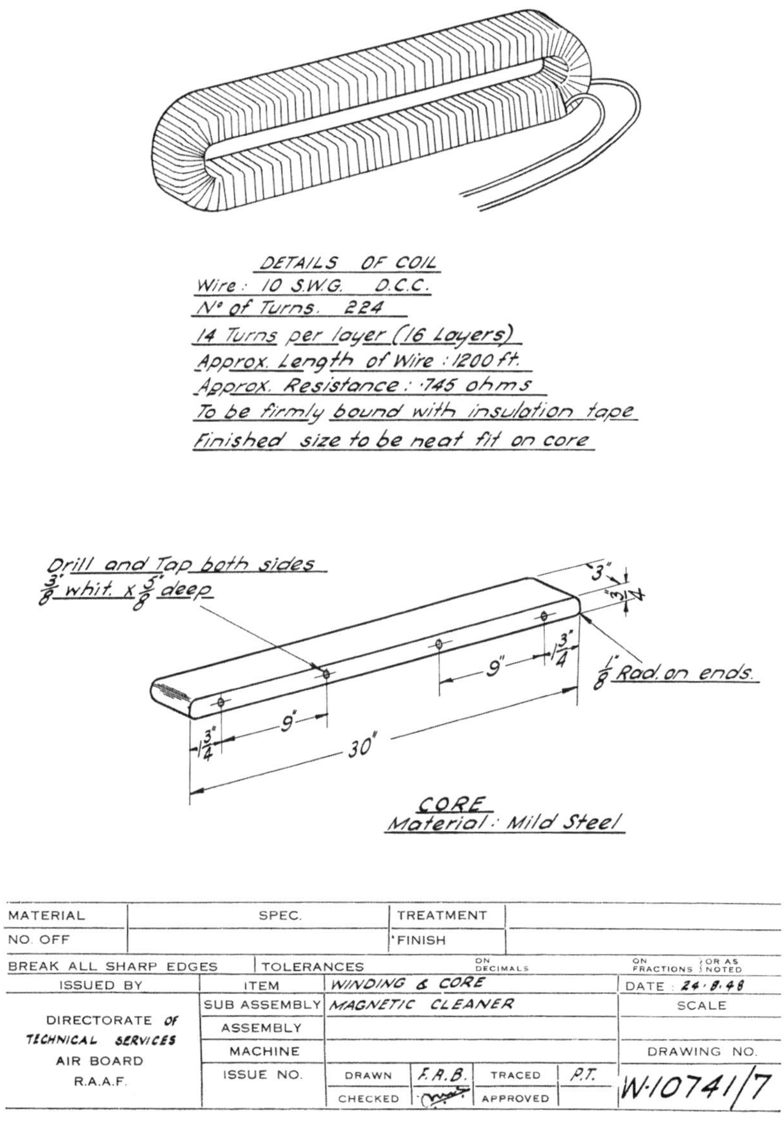

The magnetic cleaner drawings were all dated in August 1948. This is the detail of the coil winding around a central mild steel bar core.

The RAAF also mounted a water separator in a modified Australian No.4 trailer. The separator was connected between the fuel delivery tanker of a civilian supplier and the holding tanks at RAAF aerodromes to ensure the fuel delivered was free of water contamination. As they were constantly employed on aerodromes, the water separator trailers were painted bright yellow synthetic enamel with the RAAF registration across the rear panel in black. Note the guard welded to the rear underside to limit how far the trailer could be tipped rearward.

The post-war period saw 8-cwt trailers that were surplus to military requirements being transferred to other government departments. This trailer was used by the Department of Supply as a ration trailer for a lengthy reconnaissance from Woomera to Lake Christopher in 1951 during the development of the firing range for the Long Range Weapons Establishment. It was modified with a hinged rear tailgate and a spare wheel assembly mounted on the draw bar. The wheels and tyres were changed to 7.50 x 16 commercial pattern. It was towed by a 15-cwt truck and is seen here after being flipped on its side by the rough terrain. A second trailer was more extensively modified as a liquids trailer by fitting tanks to carry 227 litres of water, 272 litres of petrol and 45 litres of oil. The track width was modified to 1.64 metres to better match the towing vehicle. Vacuum air brakes and 11.00 x 16 smooth balloon tyres were fitted.

APPENDIX 1:

Mechanical Vehicle Field Specification

An MVFS was compiled by the Master General of the Ordnance (MGO) for each wheeled vehicle in service (except AFVs). Each MVFS was divided into four sections. Section 1 was general details, including the Army census number, nomenclature, a description of the vehicle's role, body type, motive power, and so on. Section 2 was a list of the tools and equipment to be issued with the vehicle, while Section 3 provided dimensional information.

Notified in G.O.'s dated, 31st July, 1944

RESTRICTED

The information given in this Document is not to be communicated, either directly or indirectly, to the Press or to any Person not authorised to receive it.

AUSTRALIAN MILITARY FORCES

MECHANICAL VEHICLE
FIELD SPECIFICATION NO.1027.

TRUCKS 1/4-TON (USA)

FORD 4X4

1944

Prepared by Staff of the Master-General of the Ordnance and issued under the Direction of the Commander-in-Chief, Headquarters, Australian Military Forces.

Lithographed by Authority:
McLaren & Co. Pty. Ltd., Melbourne

MECHANICAL VEHICLE FIELD SPECIFICATION

General Details

Army Nomenclature - Trucks, 1/4-ton (USA)

Make - Ford

Vehicle Classification - "B"

Vehicle Type - 4 x 4

Census No. (Ref Schedule "M") - 6010

Vehicle Log Book - (AAB-20B)

Type of body - Open 4 passenger with 2 individual front seats, and 1 wide rear seat.

Role - A double seated vehicle to carry personnel, primarily for reconnaissance purposes, to transport light cargo and to tow light artillery.

Wheelbase - 80"

Chassis

Make - Ford Model - GPW(USA)

Speed - Max mph - 65

Engine

Make - Ford Displacement - 134 cub. ins

No of Cycls. - 4 Bore - 3-1/8-in Stroke - 4$\frac{3}{8}$"

Brake HP - 60 at 4000 RPM

Torque - 105 ft lbs at 2000 RPM

Oil sump capacity - 3$\frac{1}{2}$-qts Grade - OE30

Cooling System

Total capacity - 9-qts

Fuel Tanks

No. fitted - 1 Capacity - 12$\frac{1}{2}$-gall

Consumption - Average Normal Range 170 miles

Gearbox

No of speeds forward - 3 Direct drive on 3rd

Lubricant Capacity - 1$\frac{3}{4}$-pts Lubricant Grade - G090

Transfer Case - (Auxiliary Gear Box)

Lubricant Capacity - 2$\frac{1}{2}$-pts Lubricant Grade - G090

Dual Ratio - Fitted in Transfer Case

Brakes

Service - Hydraulic

Parking - Hand

Front Axle

Driven - if required — Lubricant Capacity - 2 pts.

Lubricant Grade G090

Rear Axle

Number - One — Lubricant Capacity - 2 pts.

Lubricant Grade - G090

Steering

Left Hand

Chassis Lubrication

Pressure Type

Wheels	Rim Size	Rim Type
No. fitted - 4	400 - 16	Drop Centre
Spares - 1	" "	" "

Tyres

Type	Size	Pressures
Pneumatic - Non Direction-al Tread	Front - 600 - 16	35 lbs.
	Rear - 600 - 16	35 lbs.

Battery

No. of Plates - 15

Voltage - 6

Capacity - 116 Amp Hour

Overall Dimensions

Length - 10 - 13/16"

Width - 7-1/16"

Height - 9-1/4"

Lamps Sealed Beam

No. fitted -

Head - 2

Tail - 2

All capacities are shown in Imperial Measure.

Changes to the above Specification or grades of Lubricant to be used will be notified in Mechanization Circulars or other relevant publications.

TRUCKS, 1/4-TON (U.S.A.)

The following tools and equipment are supplied with the vehicle

Section	Voc. No.	Designation	Quantity	See Note
		DRIVERS MAINTENANCE KIT		
LV6-CT		Books, instruction	1	M
LV6-MT1	0828-Z	Brushes, cleaning engine 11"	1	O
LV6-MT1	0842-Z	Cans, oil lubricating 1/2-pt.	1	M
LV6-MT1	3578-Z	Dipstick, uncalibrated, Hardwood	1	O
LV6-MT1	0964-Z	Gauges, tyre pressure 10-80-lbs.	1	O
F	FA2138	Hammers, handles engineers ballpane 1-lb.	1	M
LV6-MT1	0973-Z	Injectors, lubricant	1	M
LV6-MT1	NIV	Jacks, lifting screw type (A1240)	1	M
LV6-MT2	1099-Z	Levers, tyre flat 24½"	2	O
F	A-374	Pliers, combination slip joint	1	M
LV6-MT1	1000-Z	Pourers, oil beaked 1-pt.	1	O
LV6-MT1	1001-Z	Pourers, petrol diecast with washer and strainer	1	O
LV6-MT1	1004-Z	Pumps, tyre hand	1	O
LV6-MT2	MV1424-Z	Rolls, tool vehicle	1	M
F	FAA-3621	Screwdrivers, G.S. (Aust.) 6-in	1	M
F	FA-4827	Spanners, Adj. 11"	1	M
F	FAA-3651	Spanners adj. Crescent type 8"	1	M
		Spanners D.E. -		
F	FA4894	3/8" x 7/16"	1	M
F		1/2" x 19/32"	1(a)	M
F	A-598	9/16" x 11/16"	1(a)	M
F	A-599	5/8" x 25/32"	1(a)	M
F		3/4" x 7/8"	1(a)	M
Stationery Supply		Vehicle Log Book (AAB-20B)	1	O
		Wrenches -		
LV6-MT2	MV1231-Z	Spark plug w/handle	1	M
LV6-MT2	MV1430-Z	Wheel nut and Jack Handle	1	M
"	MV403-Z	Wheel bearing nut	1	M
"	MV1427-Z	Fluted socket head screw	1	M
"	MV1426-Z	Drain plug	1	M
"	MV402-Z	Wheel hub puller	1	M
"	MV1428-Z	Hydraulic brake	1	M
		EXPENSE STORES		
LV6-MT3	1070-Z	Boxes tin spare bulbs	1	O
LV7-Wls.	GPW8620	Belts, fan spare (44 1/8" x 11/16" V42°)	1	O
		Bulbs, auto electric, spare -		
LV7	GPW-13007	Head	1	O
LV7	B-13466	Blackout	1	O
LV7	NPN	Tail	1	O
LV6-MT6	0502-Z	Cores valve tyre	1	O
LV6-MT4	1723-Z	Fuses automotive glass tube type	1	O
LV6-MT6	0526-Z	Outfits, tube repair	1	O
LV6-MT4	SPR-20	Plugs sparking 14 m.m (Spare)	1	O
W2	WB0968	Tape, insulating 3/4" (1/4-lb rolls)lbs	1/4	O
G2	GB-1813	Wire, copper soft No.20 SWG 1/4-lb rolls lbs.	1/4	O
		Equipment & fittings supplied with the vehicle		
LV6-MT13	NIV	Attachment, towing (GPW-5182)	1	M
LV7	NIV	Bars, bumper (GPW-17750)	1	M
LV6-MT13	NIV	Boxes, tool LH(GPW-1146146) RH(GPW-1146147)	2	M

TRUCKS, 1/4-TON (U.S.A.)

Equipment and fittings supplied with the vehicle (Contd.)

Section	Voc. No.	designation	Quantity	See Note ∅
LV6-MT13	NIV	Brackets holding shovels G.S. (GPW-1128258)	1	M
LV6-MT13	NIV	Bow assembly complete sets (GPW-1151268)	1	M
LV6-MT4	2320-Z	Battery 6 volt 15 plate	1	M
LV7-Wys.	A3216	Covers, canvas hood	1	M
" "	A3211	Covers, canvas windscreen	1	M
LV6-MT14	NIV	Covers, pneumatic spare	1	M
LV7-Wls.	GPW-1163206	Cushions Seats, front	2	M
" "	GPW-1163846	Cushions seats, rear	1	M
LV7-FD	GPW-1164812	Cushions, backrests front seats	2	M
" "	GPW-1166800	Cushions, backrests back seat	1	M
LV7 "	GPW-1120040 GPW-1120041	Curtains, side equipment sets	1	M
LV6-MT11	2738-Z	Chains, tyre consisting of 2 chains 1 tensioner sets	1	M
LV6-MT1	0880-Z	Extinguisher, fire filled 1-qt. w/ bracket	1	M
LV7-Wls.	GPW-8307	Guards, radiator	1	M
LV6-MT2	NIV	Handles, starting GPW-17036	1	M
LV7-Wls.	GPW-3685	Keys Ignition (sets)	1	M
LV6-MT1	3592-Z	Holders vehicle Log Book steel	1	O
LV7-Wls.		Lamps electric vehicle		
" "	GPW-13005 GPW-13006	Head w/bulb	2	M
" "	GPW-13200 GPW-13201	Blackout head with bulb	2	M
	A6142	Blackout driving with bulb	1	M
" "	GP-13404-B2	Tail and stop w/bulbs	1	M
	GP-13405-B2	Tail & Service w/bulbs	1	M
" "	21CS-17682-B123	Mirrors, driving exterior w/brackets	1	M
LV6-MT11	2057-Z	Ropes, towing steel medium	1	O
LV6-MT13	NIV	Straps web GPW-1152730	7	M
LV6-MT13	NIV	Straps safety GPW-1131414	2	M
" "	NIV	Straps holding shovels, GS GPW-1128237 GPW-1128247	2	M
LV7-FD	GPW-1160327	Seats front	2	M
"	GPW-1161326	Seats rear	1	M
J1		Shovels, G.S.	1	M
LV6-MT14	NIV	Tubes, inner pneumatic, spare	1	M
LV7-FD	GPW-17500	Wipers, windscreen hand operated	2	M
LV7-FD	GPW-1015	Wheels, spare 4.00 x 16	1	M

Items shown in this Field Specification are initial equipment only.

Maintenance will be provided in accordance with this Field Specification except where items are marked thus (a) :-

Maintenance of the items marked (a) will be provided in accordance with the nomenclature and size as set out hereunder.

Initial equipment as shown on Specification	Maintenance replacements to be	
Spanners, DE -	Spanners, DE -	
1/2 x 19/32"	1/2 x 9/16"	FA 4895
9/16 x 11/16"	No Maintenance	
5/8 x 25/32"	5/8 x 3/4"	4888
3/4 x 7/8"	7/8 x 1"	4889

NOTE Ø This column denotes the origin of initial supply of the itme thus.

O Supplied by Army Ordnance Service.

M Supplied by Vehicle Manufacturer or Assembly.

FIELD SPEC. No 1027

CLASS: TRUCKS, $\frac{1}{4}$ TON.	VEHICLE WEIGHTS	UNLADEN	LADEN	GROUND CLEARANCE	INS. 8
DETAIL: MODEL G.P. (U.S.A.)	Front Axle	·56 Tons.	·56 Tons.		
	Rear Axle	·52 Tons.	·77 Tons.	TURNING CIRCLE	FT. 36
MAKE: FORD. 4x4.	Total	1·08 Tons.	1·33 Tons.		

SHIPPING NOTES	
Shipping space standing	5·68T
Max. height cut down.	4'-5"

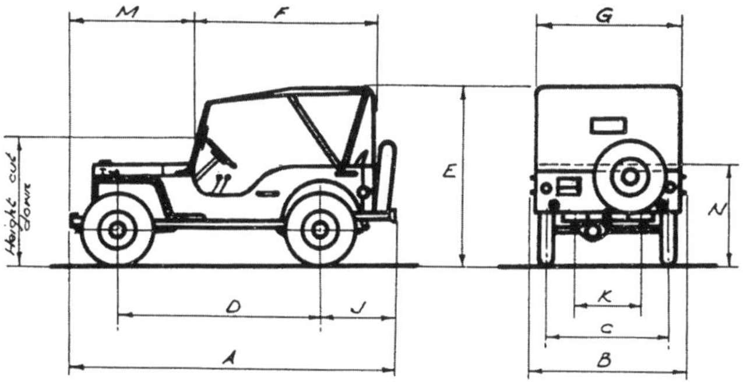

BRIDGE CLASSIFICATION: Unloaded 1 Loaded 2

RAILWAY GAUGES: Vehicle will pass all Australian Railway Loading Gauges.

A	B	C	D	E	F	G	H	J	K	L	M	N
11'-0"	5'-2"	4'-1"	80"	6'-1"	5'-8½"	4'-8½"	—	2'-7"	2-2½"	—	4'-7"	3-3½"

Tyres: 6·00 x 16

Prepared from

Notified in G.O.'s dated, 31st July, 1944

RESTRICTED

The information given in this Document is not to be communicated, either directly or indirectly, to the Press or to any Person not authorised to receive it.

AUSTRALIAN MILITARY FORCES

MECHANICAL VEHICLE FIELD SPECIFICATION NO.1028.

TRUCKS 1/4-TON (USA)

WILLYS 4X4

Superseded by 1027A 1028A } 1945.

1944

Prepared by Staff of the Master-General of the Ordnance and issued under the Direction of the Commander-in-Chief, Headquarters, Australian Military Forces.

Lithographed by Authority:
McLaren & Co. Pty. Ltd., Melbourne

MECHANICAL VEHICLE FIELD SPECIFICATION

General Details

Army Nomenclature - Trucks, 1/4-ton (USA)

Make - Willys

Vehicle Classification - "B"

Vehicle Type - 4 x 4

Census No. (Ref Schedule "M") - 6010

Vehicle Log Book - (AAB-20B)

Type of body - Open 4 passenger with 2 individual front seats and 1 wide rear seat.

Role - A double seated vehicle to carry personnel primarily for reconnaissance purposes, to transport light cargo and to tow light artillery.

Wheelbase - 80"

Chassis

Make - Willys Model - MB(USA)

Speed - Max mph - 65

Engine

Make - Willys Displacement - 134 cub ins

N o of Cyls - 4 Bore - 3-1/8" Stroke - 4$\frac{3}{8}$"

Brake HP - 60 at 4000 rpm

Torque - 105 ft lbs at 2000 rpm

Oil sump capacity - 3$\frac{1}{2}$-qts. Grade - OE30

Cooling System

Total capacity - 9-qts

Fuel Tanks

No fitted - 1 Capacity - 12-1/2-gall

Consumption - Average Normal Range 170 miles

Gearbox

No of speeds forward - 3 Direct drive on 3rd

Lubricant Capacity - 1$\frac{3}{4}$-pts Lubricant Grade G090

Transfer Case - (Auxiliary Gear Box)

Lubricant Capacity - 2-1/2-pts Lubricant Grade - G090

Dual Ratio - Fitted in Transfer Case

Brakes

Service - Hydraulic

Parking - Hand

Front Axle

Driven - if required	Lubricant Capacity 2-pts
	Lubricant Grade - GO90

Rear Axle

Number - one	Lubricant Capacity 2-pts
	Lubricant Grade - GO90

Steering

Left Hand

Chassis Lubrication

Pressure type

Wheels	Rim Size	Rim Type
No fitted - 4	400-16	Drop centre
Spares - 1		

Tyres

Type	Size	Pressures
Pneumatic Non-Directional Tread	Front 600-16	35-lbs
	Rear 600-16	35-lbs

Battery

	Overall Dimensions
No of Plates - 15	
Voltage - 6	Length - 10-13/16"
Capacity - 116 amp hr	Width - 7-1/16"
	Height - 9-1/4"

Lamps Sealed Beam

No fitted -

Head - 2

Tail - 2

All capacities are shown in Imperial Measure

Changes to the above Specification or grades of Lubricant to be used will be notified in Mechanization Circulars or other relevant publications.

Section	Voc No	Designation	Quantity	See Note
		DRIVERS MAINTENANCE KIT		
LV6-CT		Books instruction	1	M
LV6-MT1	0828-Z	Brushes cleaning engine 11"	1	O
LV6-MT1	0842-Z	Cans oil lubricating 1/2-pt	1	M
LV6-MT1	0874-Z	Dipstick fuel wooden calibrated(FD)	1	O
LV6-MT1	0964-Z	Gauges tyre pressure 10-80-lbs	1	O
F	GP17042	Hammers engineers ballpane 1-lb	1	M
LV6-MT1	GP17125	Injectors lubricant (GP17125)	1	M
LV6-MT1	GPW17080	Jacks lifting screw type	1	M
LV6-MT2	1099-Z	Levers tyre flat 24-1/2-	2	O
F	GP17028	Pliers combination slip-joint	1	M
LV6-MT1	1000-Z	Pourers oil beaked 1-pt	1	O
LV6-MT1	1001-Z	Pourers petrol diecast w/washer and strainer	1	O
LV6-MT1	1004-Z	Pumps tyre hand	1	O
LV6-MT2	MV1424-Z	Rolls tool vehicle	1	M
F	GP17020	Screwdrivers GS 6-in	1	M
F	GP17021	Spanners adjustable screw type 11"	1	M
F	GP17023	Spanners adjustable Crescent type 8"	1	M
F		Spanners DE -		
F	FA4894	3/8-in x 7/16-in	1	M
F		1/2-in x 19/32-in	1(a)	M
F	GP17045	9/16-in x 11/16-in	1(a)	M
F	GP17046	5/8-in x 25/32-in	1(a)	M
F	GP17047	3/4-in x 7/8-in	1(a)	M
Stationery Supply		Vehicle Log Book (AAB-20B)	1	O
LV6-MT2		Wrenches -		
LV6-MT2	MV1231-Z	Spark plug w/handle	1	M
LV6-MT2	MV1430-Z	Wheel nut and Jack Handle	1	M
LV6-MT2	MV403-Z	Wheel bearing nut	1	M
LV6-MT2	MV1427-Z	Fluted socket head screw	1	M
LV6-MT2	MV1426-Z	Drain plug	1	M
LV6-MT2	MV1402-Z	Wheel hub puller	1	M
LV6-MT2	MV1428-Z	Hydraulic brake	1	M
		EXPENSE STORES		
LV6-MT3	1070-Z	Boxes tin spare bulbs	1	O
LV7-FD	GPW8630	Belts fan spare	1	O
LV6-MT3		Bulbs auto electric spare -		
LV6-MT3	GPW13007	Headlamp	1	O
LV6-MT3	B13466	Blackout	1	O
LV6-MT3	NPW	Tail	1	O
LV6-MT6	0502-Z	Cores valve tyre	1	O
LV6-MT4	1723-Z	Fuses automotive glass tube type	1	O
LV6-MT6	0526-Z	Outfits tube repair	1	O
LV6-MT4	SPR-20	Plugs sparking 14 mm (spare)	1	O
W2	WB0968	Tape insulating 3/4" (1/4-lb rolls)lbs	1/4	O
G2	GB1813	Wire copper soft No 20 SWG(1/4-lb rolls) lbs	1/4	O

The following tools and equipment are supplied with the vehicle

Section	Voc No	Designation	Quantity	See Note
		EQUIPMENT AND FITTINGS SUPPLIED WITH THE VEHICLE		
LV6-MT13	N I V	Attachment towing (GPW5182)	1	M
LV6-MT13	N I V	Bars bumper (GPW17750)	1	M
LV6-MT13	N I V	Boxes tool vehicle LH(GPW1146146)	1	M
LV6-MT13	N I V	Boxes tool vehicle RH(GPW1146147)	1	M
LV6-MT13	N I V	Bracket holding shovels GS(GPW1128258)	1	M
LV6-MT13	N I V	Bow assy complete(sets)(GPW1151268)	1	M
LV6-MT4	2320-Z	Battery 6-volt 15-plate	1	M
LV7-FD		Covers canvas hood(GPW1152700)	1	M
LV7-FD		Covers canvas windscreen(GPW1103214)	1	M
LV6-MT14	N I V	Covers pneumatic spare (600 x 16)	1	M
LV7-FD		Cushions seats front (GPW1163206B)	2	M
LV7-FD		Cushions seats rear (GPW1163846B)	1	M
LV7-FD		Cushions backrest front seats(GPW-1164812B)	2	M
LV7-FD		Cushions backrest rear seats (GPW-1166800B)	1	M
LV7-FD		Curtains side equipmtne (sets) (GPW1120040) (GPW1120041)	1	M
LV6-MT11	2733-Z	Chains tyre consisting of 2 chains one tensioner (sets)	2	M
LV6-MT1	0880-Z	Extinguisher fire filled 1-qt w/bracket	1	M
LV7-FD	GPW8307	Guards radiator	1	M
LV6-MT2	N I V	Handles starting (GPW17036)	1	M
LV6-MT1	3592-Z	Holders vehicle Log Book steel	1	O
LV7-Jeep	CPW3685	Keys ignition (sets)	1	M
		Lamps electric vehicle -		
LV7-FD	GPW130-05	Head with bulb LH	1	M
LV7-FD	GPW130-06	Head with bulb RH	1	M
LV7-FD	GPW132-00	Blackout head with bulb LH	1	M
LV7-FD	GPW132-01	Blackout head with bulb RH	1	M
LV7-FD	GP134-04B	Tail and stop with bulbs	1	M
LV7-FD	GP134-05B	Tail and service with bulbs	1	M
LV7-FD	21CS1176-82B	Mirrors driving exterior with bracket	1	M
LV6-MT11	2057-Z	Ropes towing steel medium Aust.No.1	1	O
LV6-MT13	N I V	Straps web GPW-1152730	7	M
LV6-MT13	N I V	Straps safety GPW1131414	2	M
LV6-MT13	N I V	Straps holding shovel GS(GPW1138347) (GPW1128247)	2	M
LV7-FD	GPW116-0327	Seats front	2	M
LV7-FD	GPW116-1326	Seats rear	1	M
J1		Shovels GS	1	M
LV6-MT14	N I V	Tubes inner pneumatic (600 x 16)spare	1	M
LV7-FD	GP17500	Wipers windscreen hand operated	2	M
LV7-FD	GPW1015	Wheels spare (400 x 16)	1	M

Items shown in this Field Specification are initial equipment only

Maintenance will be provided in accordance with this Field Specification except where items are marked thus(a):

Maintenance of the items marked (a) will be provided in accordance with the nomenclature and size as set out hereunder:

Initial equipment as shown on Specification	Maintenance replacements to be	
Spanners DE -	Spanners DE -	
1/2" x 19/32"	1/2" x 9/16"	FA4895
9/16" x 11/16"	No Maintenance	
5/8" x 25/32"	5/8" x 3/4"	4888
3/4" x 7/8"	7/8" x 1"	4889

NOTE:-

Ø This column denotes the origin of initial supply of the items thus :-

O - Supplied by Army Ordnance Service

M - Supplied by Vehicle Manufacturer or Assembly Contractor.

FIELD SPEC. Nº1028

		Vehicle Weights	Unladen	Laden	Ground Clearance	Ins. 8	Shipping Notes	
Class:	Trucks, 1/4 ton.	Front Axle	·56 Tons.	·56 Tons.				
Detail:	Model MB (U.S.A)	Rear Axle	·52 Tons.	·77 Tons.	Turning Circle	Ft. 36	Shipping space standing	5.68T.
Make:	Willys 4x4	Total	1·08 Tons.	1·33 Tons.			Max. height cut down.	4'-5"

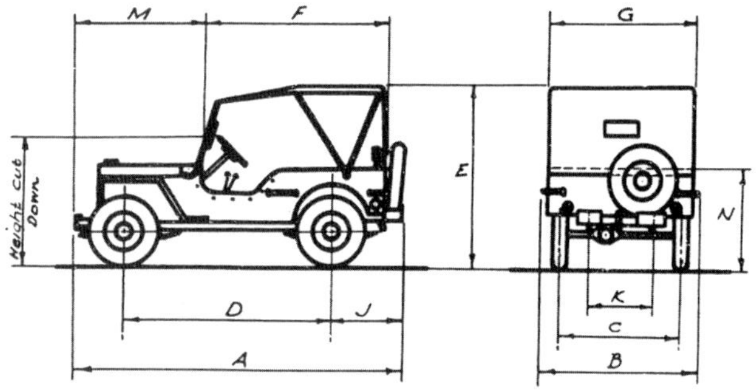

Bridge Classification: Unloaded 1 Loaded 2

A	B	C	D	E	F	G	H	J	K	L	M	N
11'-0"	5'-2"	4'-1"	80"	6'-1"	5'-8½"	4'-8½"	—	2'-7"	2'-2½"	—	4'-7"	3'-3⅜"

Railway Gauges: Vehicle will pass all Australian Railway Loading Gauges

Tyres: 6·00 x 16

Prepared from

Notified in G O s dated.30th November, 1945

RESTRICTED

The information given in this document is not to be communicated, either directly or indirectly, to the Press or to any person not authorized to receive it.

AUSTRALIAN MILITARY FORCES

MECHANICAL VEHICLE FIELD SPECIFICATION NO 1027A / 1028A

(Supersedes MVFS 1027 and 1028 dated 31/7/44).

TRUCKS 1/4-TON GS (USA).

FORD 4X4

WILLYS 4X4.

1945 Specification

Prepared by Staff of the Master General of the Ordnance and issued under the Direction of the Commander-in-Chief, Headquarters, Australian Military Forces.

Lithographed by Authority -

Field Specification No 1027A.

Field Specification No 1028A.

Supersedes MVFS 1027 & 1028 Dated 31/7/44

MECHANICAL VEHICLE FIELD SPECIFICATION

General Details

Army Nomenclature - Trucks 1/4-ton (USA).

Make - Ford or Willys

Vehicle Classification - "B"

Vehicle Type - 4 x 4

Census No (Ref Schedule "M") 6010

Vehicle Log Book (AAB-20B)

Type of Body - Open type passenger body with folding hood. Spare wheel is mounted on rear of body and pintle hook is provided to haul trailed loads.

Role - A general purpose vehicle used for transporting personnel and towing trailers 8-cwt GS.

Wheelbase - 80"

Chassis

Make - Ford or Willys Model GPW or MB

Speed - Max mph - 65

Engine

Make - Ford Displacement - 134 cub ins

No of Cyls - 4 Bore - 3-1/8" Stroke 4-3/8"

Brake HP 60 at 3600 rpm

Torque - 105 ft lbs at 2000rpm

Oil sump capacity 3-1/2-qts Grade - OE30

Cooling System

Total capacity - 9-1/4-qts

Fuel Tanks

No fitted - 1 Capacity - 12-1/2-galls

Consumption - Average Normal Range 170 miles

Gear box

No of speeds forward - 3 Direct drive on 3rd

Lubricant Capacity 1-pt Lubricant Grade G090

Transfer Case (Auxiliary Gearbox)

Lubricant Capacity 2-1/2-pts   Lubricant Grade GO90

Dual Ratio - Fitted in TransferCase

Brakes

Service - Hydraulic

Parking - Hand

Front Axle

Driven - if required   Lubricant Capacity 2-pts

Lubricant Grade GO90

Rear Axle

Number - One   Lubricant Capacity 2-pts

Lubricant Grade GO90

Steering

Left Hand

Chassis Lubrication

Pressure Type

Wheels	Rim Size	Rim Type
No fitted - 4	16"	Drop Centre
Spares - 1		

Tyres

Type	Size	Pressures
Pneumatic Non-Directional Tread	Front 600 x 16	Normal Road 30-lbs
		Cross Country 20-lbs
	Rear 600 x 16	Normal Road 30-lbs.
		Cross Country 20-lbs

Battery

No of Plates - 15

Voltage - 6

Capacity - 116 amp hrs

Overall Dimensions

Length - 10"

Width - 7"

Height - 8-5/16"

Lamps Prefocussed

No fitted :-

Head : 2

Blackout Driving - 1

Tail and Stop - 2

Blackout Head - 2

Applicable Publications

Trucks 1/4-ton 4 x 4 - FORD

Maintenance Manual TM10-1101, 1941
Maintenance Manual TM10-1349, Sept 1942
Technical Manual TM9-803, Feb 1944
Instruction Book Driving & Maintenance 31 July 1944
Parts List TM10-1100 Oct 1941
Parts List TM10-1348 Feb 1943

Trucks 1/4-ton 4 x 4 - WILLYS

Maintenance Manual TM10-1207 Aug 1941
Maintenance Manual TM10-1513 Jan 1943
Technical Manual TM9-803 Feb 1944
Instruction Book Driving & Maintenance 31 July 1944
Parts List TM10-1206 Feb 1942
Parts List TM10-1512 Sept 1942

All capacities are shown in Imperial Measure.

Changes to the above Specification on Tyre Inflation Pressures or Grades of Lubricant to be used will be notified in MGO Equipment Orders or other relevant publications.

Field Specification No 1027A.

Field Specification No 1028A.

Supersedes MVFS 1027 & 1028 Dated 31/7/44

The following tools and equipment are supplied with the vehicle

Section	Vocab No	Designation	Quantity	See Note ø
		DRIVERS MAINTENANCE KIT		
LV6-CT	N I V	Books instruction	1	M
LV6-MT1	0828-Z	Brushes cleaning engine 11-in	1	O
LV6-MT1	0842-Z	Cans oil lubricating	1	M
LV6-MT1	0872-Z	Dipstick fuel wooden calibrated	1	O
LV6-MT1	0964-Z	Gauges tyre pressure 10-80-lbs	1	M
LV6-MT2	MV1062-Z	Gauges spark plug gap universal	1	O
F	FA2138	Hammers handled engineers ball-pane 1-lb	1	M
LV6-MT1	0974-Z	Injectors lubricant pistol grip 10-oz	1	M
LV6-MT1	0983-Z	Jack lifting screw type 1-1/2-ton	1	M
LV6-MT2	1099-Z	Levers tyre flat 24-1/2-in	2	O
LV6-MT1	0998-Z	Plates Base Jacks wooden	1	O
F	FAA3017	Pliers combination and side cutting 7-in (Aust) prs	1	M
LV6-MT1	1000-Z	Pourers oil beaked 1-pt	1	O
LV6-MT1	1001-Z	Pourers petrol die cast with washer and strainer	1	O
LV6-MT2	MV402-Z	Pullers front hub	1	M
LV6-MT1	1004-Z	Pumps tyre hand	1	M
LV6-MT2	MV1424-Z	Rolls tool vehicle leather	1	M
F	FAA3629	Screwdrivers GS (Aust) through tanged round shank 12-in	1	M
F		Spanners AMF - DE -		
	FAA3659	3/8-in x 7/16-in	1	M
		1/2-in x 19/32-in	1(a)	M
		9/16-in x 11/16-in	1(a)	M
		5/8-in x 25/32-in	1(a)	M
	FAA3663	3/4-in x 7/8-in	1	M
F	FA4827	Spanners adjustable 11-in	1	M
LV6-MT2	MV1194-Z	Tools tyre valve repair	1	O
Stationery Supply		Vehicle Log Book (AAB-20B)	1	O
LV6-MT2	MV1231-Z	Wrenches spark plug with handle	1	M
LV6-MT2	MV1426-Z	Wrenches drain plug	1	M
LV6-MT2	MV1427-Z	Wrenches fluted socket head	1	M
LV6-MT2	MV1428-Z	Wrenches hydraulic brake bleeder screw	1	M
LV6-MT2	MV1429-Z	Wrenches wheel hub	1	M
LV6-MT2	MV1430-Z	Wrenches wheel nuts and jack handle	1	M
		EXPENSE - STORES.		
LV6-MT3	1070-Z	Boxes tin spare bulb	1	O
LV7/Jeep	GPW8620	Belts fan spare (V9)	1	M
LV6-MT3	17-L-5215	Bulbs auto electric spare - 6-8 volt, 3 cp	1	M
LV6-MT6	0502-Z	Cores valve tyre (tins of five) tins	1	M

The following tools and equipment are supplied with the vehicle

Section	Vocab No	Designation	Quantity	See Note
		EXPENSE STORES (CONTD)		
LV6-MT6	0493-Z	Caps valve tyre (tins of 5) tins	1	M
LV7-JEEP		Lamps electric sealed beam units		
	GP13404-B	Blackout stop 6-8 volt 3 cp	1	M
	GP13405-B	Blackout tail 6-8 volt 3 cp	1	M
	GPW13491A) GPW13494)	Service tail & stop 6-8 volt 21-3 cp	1	M
LV6-MT6	0526-Z	Outfits tube repair	1	O
LV6-MT4	SPR-20	Plugs sparking spare	1	M
G1		Pins keep split assorted pkts	1	M
W2	WB0968	Tape insulating 3/4"(1/4-lb rolls) lbs	1/4	M
G2	GB1813	Wire copper soft No.20 SWG (1/4-lb rolls) lbs	1/4	M
		EQUIPMENT SUPPLIED WITH VEHICLE		
LV7/FD	GPW1152700	Covers canvas hood	1	M
LV6-MT14		Covers pneumatic spare 600 x 16	1	M
LV6-MT11	2738-Z	Chains tyre non-skid (sets of 4) sets	1	M
LV7/JEEP	GPW1120040 GPW1120041	Curtains side equipment sets	1	M
LV6-MT1	0880-Z	Extinguishers fire filled 1-qt tetrachloride type	1	M
LV6-MT2	MV1425-Z	Handles starting	1	M
LV6-MT11	4228-Z	Ropes towing steel light (Aust) No.3	1	O
J1	JAA3664	Shovels GS (Aust)	1	M
LV6-MT14		Tubes inner pneumatic spare 600 x 16	1	M
LV7/JEEP	GPW1015	Wheels spare 400 x 16	1	M

Items shown in this Field Specification are initial equipment only.

Maintenance will be provided in accordance with this Field Specification except where items are marked thus (a):-

·Maintenance of items marked (a) will be provided in accordance with nomenclature and size as set out hereunder:-

Initial equipment as shown in Specification		Maintenance replacements to be
1/2" x 19/32"	FAA 3661	1/2" x 9/16"
9/16" x 11/16"	FAA 3662	5/8" x 11/16"
5/8" x 25/32"		

NOTE ∅ - This column denotes source of initial supply:

(C) Supplied by Army Ordnance Service

(M) Supplied by vehicle manufacturer or assembly contractor.

FIELD SPEC. No. 1027A and 1028A.

CLASS:	TRUCKS, 1/4 TON. - U.S.A.		VEHICLE WEIGHTS	UNLADEN	LADEN	GROUND CLEARANCE	INS. 8	SHIPPING NOTES	
DETAIL:	MODEL G.P	MODEL MB	Front Axle	·56 Tons.	·56 Tons.			Shipping space standing	5·68T
MAKE:	FORD. 4×4.	WILLYS 4×4	Rear Axle	·52 Tons.	·77 Tons.	TURNING CIRCLE	FT. 36	Max. height cut down	4'-5"
			Total	1·08 Tons.	1·33 Tons.				

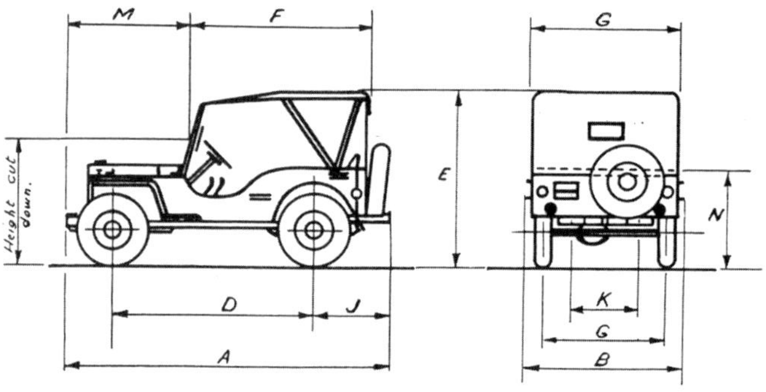

BRIDGE CLASSIFICATION: Class 2

RAILWAY GAUGES: Vehicle will pass all Australian Railway Loading Gauges.

A	B	C	D	E	F	G	H	J	K	L	M	N
11'-0"	5'-2"	4'-1"	80"	6'-1"	5'-8½"	4'-8½"	—	2'-7"	2'-2½"	—	4'-7"	3'-3½"

Tyres: 6·00 × 16

Prepared from

M.V.F.S. 1027A and 1028A.

Notified in G O s dated 30th November, 1945

> RESTRICTED
>
> The information given in this document is not to be communicated, either directly or indirectly, to the Press or to any person not authorized to receive it.

AUSTRALIAN MILITARY FORCES

MECHANICAL VEHICLE FIELD SPECIFICATION NO 1029.

TRUCKS 1/4-TON AMPHIBIOUS (USA).

FORD 4 X 4.

1945 Specification

Prepared by Staff of the Master General of the Ordnance and issued under the Direction of the Commander-in-Chief, Headquarters, Australian Military Forces.

Lithographed by Authority -

Tyres

Type	Size		Pressures	
Pneumatic Non-Directional Tread	Front	600 x 16	Normal Road	30-lbs
			Cross Country	20-lbs
	Rear	600 x 16	Normal Road	30-lbs
			Cross Country	20-lbs

Batteries

Number fitted - 2 (connected in series for 12 volts)

No of Plates - 15

Voltage - 6

Capacity - 116 amp hrs

Overall Dimensions

Length - 10"

Width - 7"

Height - 8-5/16"

Lamps Prefocussed :

No fitted :

Head : 2

Blackout Marker - 2

Blackout Driving - 1

Tail : 2

Model		Applicable Publications
1942	-	Parts List TM10-1264 October '42 Maintenance Manual TM10-1263 - 1943.

WINCH - A winch is mounted on the front of the hull and is powered by the engine through a special power take-off unit.

All capacities are shown in Imperial Measure.

Changes to the above Specification on Grades of Lubricant to be used will be notified in MGO Equipment Orders or other relevant publications.

The following tools and equipment are supplied with the vehicle

Section	Vocab No	Designation	Quantity	See Note
		DRIVERS MAINTENANCE KIT		
LV6-MT2	MV1013-Z	Bars tommy for wrenches spark plug	1	M
LV6-CT	N I V	Books instruction	1	M
LV6-MT1	0828-Z	Brushes cleaning engine 11-in	1	O
LV6-MT1	0842-Z	Cans oil lubricating	1	M
LV6-MT1	0872-Z	Dipstick fuel wooden calibrated	1	O
F	FA1720	Files smooth half round 8-in	1	O
F	FA1802	Handles medium	1	O
LV6-MT1	0895-Z	Funnels petrol 6" dia 9" spout	1	O
LV6-MT1		Gauges tyre pressure universal 10-80-lbs	1	M
LV6-MT2	MV1062-Z	Gauges spark plug gap universal	1	O
F	FA	Hammers handled engineers ballpane 1-lbs	1	M
LV6-MT1	0974-Z	Injectors lubricant pistol grip 10-oz with adaptor	1	M
LV6-MT1	0984-Z	Jack lifting screw type 3-ton	1	M
LV6-MT1	1051-Z	Lamps inspection portable with 25-ft lead	1	M
LV6-MT2	1099-Z	Levers tyre flat 24-1/2-in	2	O
LV6-MT1	0998-Z	Plates base jacks wooden	1	M
F	FAA3016	Pliers combination and side cutting 6-in (Aust) prs	1	M
LV6-MT1	1000-Z	Pourers oil beaked 1-pt	1	O
LV6-MT1	1001-Z	Pourers petrol die cast with washer and strainer	1	O
LV6-MT1	1004-Z	Pumps tyre hand operated	1	M
LV6-MT2	MV402-Z	Pullers front hub	1	M
LV6-MT2	MV1424-Z	Rolls tool vehicle leather	1	M
F	FAA3621	Screwdrivers GS(Aust) 6-in	1	M
LV6-MT1	1018-Z	Syringes oil with flexible nozzle	1	O
F		Spanners ANF - DE (Aust) -		
	FAA3659	3/8-in x 7/16-in	1	M
		1/2-in x 19/32-in	1(a)	M
		9/16-in x 11/16-in	1(a)	M
		5/8-in x 25/32-in	1(a)	M
	FAA3663	3/4-in x 7/8-in	1	M
F	FAA3651	Spanners adjustable crescent type 8-in (Aust)	1	M
F	FA4827	Spanners adjustable 11-in	1	M
LV6-MT2	MV1194-Z	Tools tyre valve repair	1	M
Stationery Supply		Vehicle Log Book (AAB-20B)	1	O
LV6-MT2	MV1231	Wrenches spark plug DE Tubular	1	M
LV6-MT2	MV1430	Wrenches wheel nut and jack handle	1	M
LV6-MT2	MV1426-Z	Wrenches drain plug	1	M
LV6-MT2	MV1428-Z	Wrenches hydraulic brake bleeder screw	1	M
LV6-MT2	MV1429-Z	Wrenches wheel hub (GPA17033)	1	M
LV6-MT2	MV1427-Z	Wrenches fluted socked head	1	M
LV6-MT2		Wrenches hull plugs (GPA17034)		

The following tools and equipment are supplied with the vehicle

Section	Vocab No	Designation	Quan-tity	See Note ø
		EXPENSE - STORES		
LV6-MT3	1070-Z	Boxes tin spare bulb	1	O
LV7-Jeep		Belts fan spare (V9)	1	M
LV6-MT3		Bulbs auto electric spare -		
LV6-MT3	1035-Z	Lamps inspection	1	M
LV6-MT3		Blackout marker	1	M
LV6-MT6	0502-Z	Cores valve tyre	1	M
LV6-MT6	0526-Z	Outfits tube repair	1	M
G1		Pins keep split assorted pkts	1	M
LV6-MT4	SPR-20	Plugs sparking spare	1	M
W2	WB0968	Tape insulating 3/4-in (1/4-lb rolls) lbs	1/4	M
G2	GB1813	Wire copper soft No.20 SWG(1/4-lb rolls) lbs	1/4	M
		EQUIPMENT SUPPLIED WITH THE VEHILCE		
J1	JA0243	Buckets water canvas Mk.V	1	M
E3	EC0023	Blocks tackle GS cordage single 2-1/2-in circ	1	M
H2	HBA0428	Cordage sisal hawserlaid, rotproof-ed, 2-1/2-in circ (Fthms)	15	M
LV6-MT11	2738-Z	Chains tyre non-skid 600 x 16(sets)	1	M
LV6-MT14		Covers pneumatic spare 600 x 16	1	M
LV6-MT13		Covers canvas body waterproof	1	M
LV6-MT13		Containers curtains empty	1	M
LV6-MT13		Curtains side equipment	2	M
LV6-MT1	0880-Z	Extinguishers fire filled 1-qt tetrachloride type	1	M
E4		Hooks boat 9-ft	1	M
LV6-MT2		Handles starting	1	M
AAMC	SUPPLY	Outfits first aid MT Vehicle	1	M
G1	GAA3005	Padlocks GS 1-1/2" with 2 keys (Aust)	2	M
E4		Pumps bilge hand	1	M
LV6-MT11	4229-Z	Ropes towing steel medium(Aust) No.3	1	M
LV6-MT11	2941-Z	Tensioners chains non-skid	1	M
LV6-MT14		Tubes inner pneumatic spare 600x16	1	M
		Sponge (GPA 1752982)	1	M
J1	JAA3664	Shovels GS (Aust)	1	M
LV7		Wheels spare 16"		

Items shown in this Field Specification are initial equipment only.

Maintenance will be provided in accordance with this Field Specification except where items are marked (a):-

Maintenance of items marked (a) will be provided in accordance with nomenclature and size as set out hereunder:-

Initial equipment as shown in specification		Maintenance replacements to be
Spanners ANF - DE -		Spanners ANF - DE -
1/2-in x 19/32-in	FAA3661	1/2-in x 9/16-in
9/16-in x 11/16-in	FAA3662	5/8-in x 11/16-in
5/8-in x 25/32-in	- -	- - - -

NOTE:- ∅ Source of initial supply :

(O) Supplied by Army Ordnance Service

(M) Supplied by vehicle manufacturer or Assembly contractor.

FIELD SPEC. Nº 1029

CLASS:	TRUCKS ¼ TON.
DETAIL:	AMPHIBIOUS (U.S.A)
MAKE:	FORD 4×4

VEHICLE WEIGHTS	UNLADEN	LADEN	GROUND CLEARANCE	INS.	SHIPPING NOTES	
Front Axle	·65 Tons.	·75 Tons.			Shipping space standing	8·2 T.
Rear Axle	·95 Tons.	1·25 Tons.	TURNING CIRCLE	FT. 36½ (LAND)	Max. height cut down	49"
Total	1·60 Tons.	* 2·00 Tons.				

* with 800 lbs. Payload.

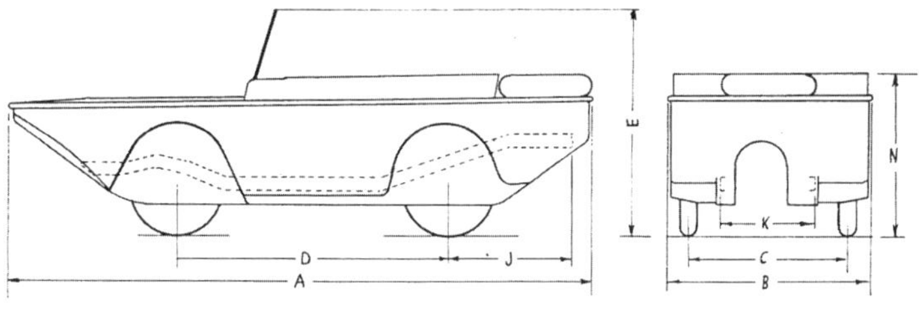

BRIDGE CLASSIFICATION: CLASS 2.

RAILWAY GAUGES: Vehicle will pass all Australian Railway Loading Gauges.

A	B	C	D	E	F	G	H	J	K	L	M	N
182"	64"	49"	84"	68"	-	-	-	39"	29"	-	-	49"

Tyres: 6·00 × 16

A.V.F.S. Nº 1029.

Notified in G.O.'s dated..................

RESTRICTED

The information given in this Document is not to be communicated, either directly or indirectly, to the Press or to any Person not authorised to receive it.

AUSTRALIAN MILITARY FORCES

MECHANICAL VEHICLE FIELD SPECIFICATION NO. 1721.

TRAILERS 2-WH 8-CWT (AUST) NOS 3 & 3A

1944

Prepared by Staff of the Master-General of the Ordnance and issued under the Direction of the Commander-in-Chief, Headquarters, Australian Military Forces.

Lithographed by Authority -

 Field Specification No.1721

MECHANICAL VEHICLE FIELD SPECIFICATION

General Details

Army Nomenclature - Trailers, 2-wh. 8-cwt., (Aust) Nos 3 & 3A

Vehicle Classification - "B"

Census No (Ref Schedule "M") - 6015

Type of Body -
Watertight steel body of welded construction mounted on channel iron chassis with spring towing eye and hinged parking supports.

Role -
For general service use by attachment to suitable Army vehicles.

Brakes -
Hand operated - only fitted to (Aust.) 3A

Wheels -

No fitted - 2

Rim Size - 16"

Tyres

Type - Pneumatic Bar Tread

Size - 6.00 x 16

TOOLS & EQUIPMENT SUPPLIED WITH THE VEHICLE

Section	Voc No.	Designation	Qty.
LV6-MT13		Covers, canvas, waterproof	1
LV6-MT2		Spanner, wheel nut and hub	1

CLASS: TRAILERS, 2 WH.	VEHICLE WEIGHTS	UNLADEN	LADEN	GROUND CLEARANCE	INS. —	SHIPPING NOTES
DETAIL: 8CWT (AUST) Nº3.	Front Axle	— Tons.	— Tons.	TURNING CIRCLE	FT. —	Shipping space standing: 4 Tons
MAKE:	Rear Axle	— Tons.	— Tons.			Max. height cut down: 3'-6"
	Total	·25 Tons.	·65 Tons.			

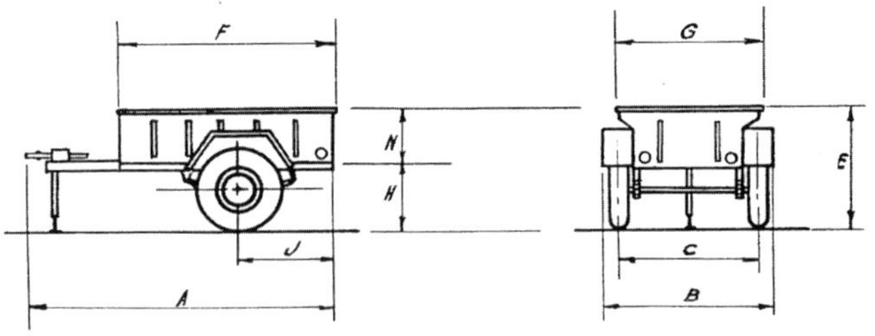

BRIDGE CLASSIFICATION: Class 1

RAILWAY GAUGES: Vehicle will pass all Australian Railway Loading Gauges.

A	B	C	D	E	F	G	H	J	K	L	M	N
8'-11½"	4'-10"	49"	—	3'-6⅜"	6'-2⅜"	4'-0⅜"	2'-0⅜"	2'-7"	—	—	—	1'-6"

Nº 1721

Tyres: 600×16

Prepared from ADD(V)3220

CLASS: TRAILERS 2 WHEEL	VEHICLE WEIGHTS	UNLADEN	LADEN	GROUND CLEARANCE	INS. —	SHIPPING NOTES
DETAIL: 8CWT (AUST.) Nº 3 A	Front Axle	— Tons.	— Tons.	TURNING CIRCLE	FT. —	Shipping space standing: 4 TONS.
MAKE:	Rear Axle	— Tons.	— Tons.			Max. height cut down: 3'-6"
	Total	·25 Tons.	·65 Tons.			

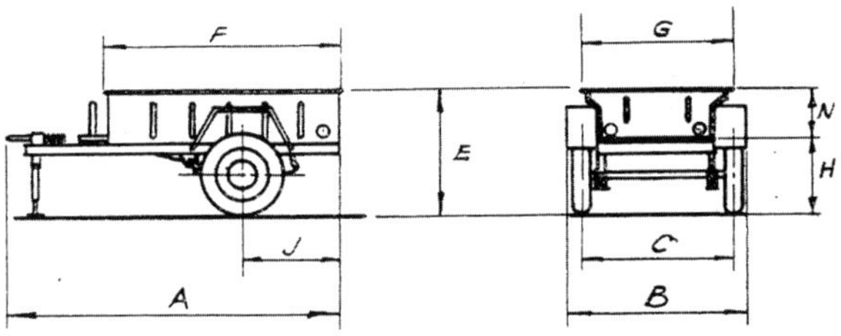

BRIDGE CLASSIFICATION: Class 1.

RAILWAY GAUGES: Vehicle will pass all Australian Railway Loading Gauges

A	B	C	D	E	F	G	H	J	K	L	M	N
8'-11½"	4'-10"	49"	—	3'-6"	6'-2¾"	4'-0⅝"	2'-0"	2'-7"	—	—	—	1'-6"

Nº 1721

Tyres: 6·00 × 16

Prepared from ADD(V)15-100

Notified in G.O.'s dated 30th November, 1944.

RESTRICTED

The information given in this Document is not to be communicated, either directly or indirectly, to the Press or to any Person not authorised to receive it.

AUSTRALIAN MILITARY FORCES

MECHANICAL VEHICLE FIELD SPECIFICATION NO. 1726.

TRAILERS 2-WH 8-CWT (U S A)

1944

Prepared by Staff of the Master-General of the Ordnance and issued under the Direction of the Commander-in-Chief, Headquarters, Australian Military Forces.

Photo-lithographed by Authority:
McCarron, Bird & Co., Printers, 479 Collins Street, Melbourne

Field Specification No 1842

MECHANICAL VEHICLE FIELD SPECIFICATION

General Details

Army Nomenclature - Kits Recovery

Role - For fitment to Trucks 1/4-ton 4x4

Army Drawing No - ADD(V) 3267

Vocabulary Section for Kit - LV6-MT13

Vocabulary No for Kit - 3228-Z

EQUIPMENT, FITTINGS AND STORES SUPPLIED WITH KIT

Vocab Section	Vocab Nos	No	Items	Qty	See Note ∅
LV6-MT13	3229Z	1	Shaft winch end	1	M
LV6-MT13	3230Z	2	Shaft handle end	1	M
LV6-MT13	3231Z	3	Support shaft	1	M
LV6-MT13	3232-Z	4	Channel	2	M
LV6-MT13	3233Z	5	Crossmember	1	M
LV6-MT13	3234Z	6	Brace,shaft support	1	M
LV6-MT13	3235-Z	7	Brace frame	1	M
LV6-MT13	3236Z	8	Brace, frame	1	M
LV6-MT13	3237-Z	9	Retainer handle	1	M
LV6-MT13	3238-Z	10	Pin	1	M
LV6-MT13	3239-Z	11	Nut handle retainer	1	M
G1	N I V	12	Rivet 3/8" dia C'sk 1-1/2" long	1	M
G1	N I V	13	Rivet 3/8" dia C'sk 2-1/4" long	1	M
G1	GAA5510	14	Bolt with nut BSW Steel Hex Hd 3/8" x 1"	4	M
G1	GA1783	15	Washer spring single turn 3/8"	4	M
G1	N I V	16	Bolt with nut BSW Hex Hd 5/8" x 8-3/4"	2	M
G1	GA0208	17	Bolt with nut BSW Steel Hex Hd 5/8" x 5"	2	M
G1	GA1787	18	Washers spring single turn 5/8"	4	M
G1	GAA5512	19	Bolt with nut BSW Steel Hex Hd 7/16" x 1"	2	M
G1	GAA5513	20	Bolt with nut BSW Steel Hex Hd 7/16" x 1-1/2"	2	M
G1	GA1784	21	Washers spring single turn 7/16"	4	M
	N I V	22	Loop 1/8" dia MS wire 1-1/2"long	2	M
	N I V	23	Spacer 1" dia MS pipe 3-5/8" long	2	M
G1	N I V	24	Chain No 10 SWG "Tenso" R1.047, 6" long	2	M
E3	N I V	25/34	Winches dawn hand 2-ton	1	M

Note ∅- This column denotes source of initial supply

O - Supplied by Army Ordnance Service
M - Supplied by vehicle manufacturer or Assembly Contractor

KITS RECOVERY FOR TRUCKS $\frac{1}{4}$ TON 4 x 4

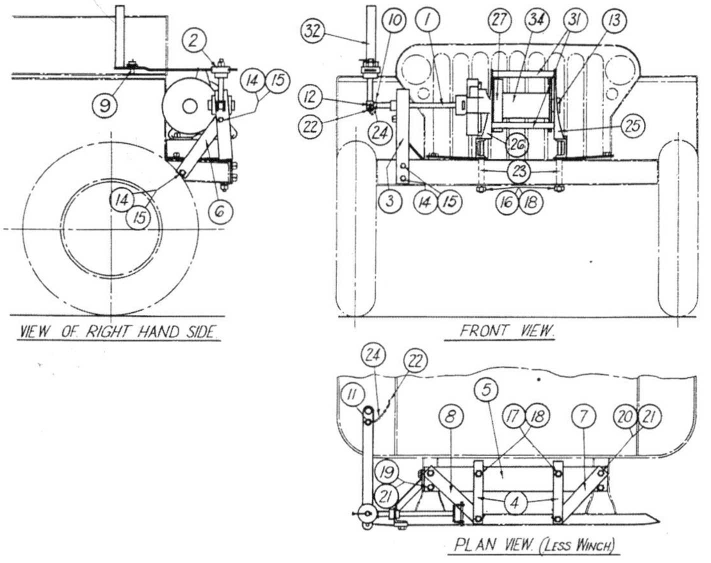

VIEW OF RIGHT HAND SIDE

FRONT VIEW.

PLAN VIEW (LESS WINCH)

PREPARED FROM (INSTALLATION DIAGRAM) A.D.D.(V.) 3267.	FIELD SPEC No. 1842

Notified in G.O.'s dated 30th November, 1944.

RESTRICTED.

The information given in this Document is not to be communicated, either directly or indirectly, to the Press or to any Person not authorized to receive it.

AUSTRALIAN MILITARY FORCES.

MECHANICAL VEHICLE FIELD SPECIFICATION NO. 1844.

KITS WELDING

FOR FITMENT TO TRUCKS 1/4-TON 4 x 4

1944.

Prepared by Staff of the Master-General of the Ordnance and issued under the Direction of the Commander-in-Chief, Headquarters, Australian Military Forces.

By Authority: K. M. STEVENSON, Government Printer, Adelaide.

MECHANICAL VEHICLE FIELD SPECIFICATION.

General Details.

Army Nomenclature—Kits, welding.

Role—For fitment to trucks, $\frac{1}{4}$ ton, 4 x 4.

Army Drawing No.—ADD(V)3299

Vocab Section for Kits—LV6-MT13.

Vocab No. for Kits—3753Z.

EQUIPMENT FITTINGS AND STORES SUPPLIED WITH KIT.

Vocab Section.	Vocab No.	No. ID.	Designation.	Quantity.	See Note.*
—	—	1	Main bearing assembly	1	M
LV6-MT13	3888-Z	2	Power take-off assembly	1	M
LV6-MT13	3886-Z	3	Base plate generator assembly	1	M
LV6-MT13	3879-Z	4	Levers operating PTO assembly	1	M
LV6-MT12	3915-Z	5	Valves, throttle assembly	1	M
LV6-MT13	3884-Z	6	Links, throttle control assembly	1	M
LV6-MT12	3907-Z	7	Governor, assembly	1	M
LV6-MT13	3874-Z	8	Brackets, governor support	1	M
LV6-MT13	3906-Z	9	Supports, switchboard RH	1	M
LV6-MT13	3905-Z	10	Supports, switchboard LH	1	M
LV6-MT13	3876-Z	11	Flywheel, generator, $15\frac{3}{4}$in. diameter	1	M
LV6-MT13	3877-Z	12	Studs, flywheel, BSW, $\frac{1}{2}$in. x $4\frac{3}{4}$in.	1	M
LV6-MT13	—	13	Pulley	1	M
LV6-MT13	3878-Z	14	Guards, governor pulley	1	M
LV6-MT13	3903-Z	15	Shaft, transmission assembly	1	M
—	—	16	Stud, carburettor	2	M
LV6-MT13	3885-Z	17	Springs, throttle control	1	M
—	—	18	Screw, adjusting	1	M
LV6-MT13	3901-Z	19	Pulleys, fan	1	M
LV6-MT13	3875-Z	20	Covers, generator canvas, 5ft. 4in. square	1	M
—	—	21	Plates, instruction	1	M
G1	GAA6407	22	Screws, BSW, round head, MS, $\frac{1}{4}$in. x $1\frac{1}{2}$in.	4	M
LV6-MT6	N I V	23	Washers, fibre, $\frac{1}{4}$in. x $\frac{1}{8}$in. thick	4	M
G1	GA1781	24	Washers, spring, single turn, $\frac{1}{4}$in.	15	M
G1	GA0700	25	Nuts, BSW, hexagonal, $\frac{1}{4}$in.	13	M
G1	GA0190	26	Bolts, BSW, hexagonal head, $\frac{1}{2}$in. x 2in., with nuts	4	M
G1	GA1785	27	Washers, spring, single turn, $\frac{1}{2}$in.	9	M
G1	GAA5525	28	Bolts, BSW, steel, hexagonal head, $\frac{1}{2}$in. x $1\frac{1}{4}$in.	2 (a)	M
G1	GAA5525	29	Bolts, BSW, steel, hexagonal head, $\frac{1}{2}$in. x 1in.	2 (a)	M
G1	GAA5570	30	Bolts, BSW, steel, bright, hexagonal head, $\frac{3}{8}$in. x $1\frac{1}{4}$in.	6	M

Equipment, Fittings, and Stores Supplied with Kit—continued.

Vocab Section.	Vocab No.	No. ID.	Designation.	Quantity.	See Note.*
G1	GA1783	31	Washers, spring, single turn, $\frac{3}{8}$in. diameter	16 (b)	M
G2	N I V	32	Keys, steel, $\frac{1}{2}$in. x $\frac{3}{8}$in. x 2in.	1	M
G1	GA1740	33	Washers, bolt or rivet, $\frac{1}{2}$in.	1	M
G1	GA0702	34	Nuts, BSW, steel, hexagonal, $\frac{1}{2}$in.	1	M
G1	GAA6405	35	Screws, BSW, MS, round head, $\frac{1}{4}$in. x $\frac{3}{4}$in.	7	M
G1	N I V	36	Bolts, SAE, MS, hexagonal head, $\frac{3}{8}$in. x $1\frac{1}{8}$in. .	8 (c)	M
G1	GAA6410	37	Screws, BSW, MS, round head, $\frac{5}{16}$in. x $\frac{3}{4}$in.	7 (a)	M
G1	GA0699	38	Nuts, BSW, steel, hexagonal, $\frac{5}{16}$in.	9	M
G1	GA1782	39	Washers, spring, single turn, $\frac{5}{16}$in. diameter	9 (d)	M
G1	GA0701	40	Nuts, ANF, steel, hexagonal, $\frac{3}{8}$in.	2	M
G1	GAA6528	41	Sets screws, BSW, hexagonal head, MS, $\frac{1}{4}$in. x $1\frac{1}{2}$in.	4 (a)	M
G1	GAA6190	42	Screws, BA, round head, MS, No. 5, $\frac{3}{8}$in. long..	4(a)	M
G1	GAA2732	43	Nuts, BA, steel, hexagonal, No. 5	4	M
G1	GAA4495	44	Washers, spring, single turn, $\frac{1}{8}$in. diameter	4	M
LV6-MT1	4000-Z	45	Belts, V, Dunlop B56 or equivalent	4	M
LV6-MT1	3999-Z	46	Belts, V, Goodyear M61 or equivalent	1	M
G1	GAA6411	47	Screws, BSW, round head, MS, $\frac{5}{16}$in. x $1\frac{1}{2}$in. ..	2	M
F	N I V	48	Generators complete with attachable switchboard	1 (e)	M
G1	GAA0283	49	Bolts, ANF, steel, bright, hexagonal head, $\frac{5}{16}$in. x 1in.	8	M

(a) For items marked thus (a) maintenance replacements will be in accordance with Vocab Nos. as shown.

(b) 8 only required when drawing ADD(V)35-163 is used.

(c) Not required when drawing ADD(V)35-163 is used.

(d) 11 required when drawing ADD(V)35-163 is used.

(e) Required only when drawing ADD(V)35-163 is used.

NOTE :—* This column denotes source of initial supply of the items thus :—

O Supplied by Army Ordnance Service.

M Supplied by vehicle manufacturer or assembly contractor.

ID Numbers in this column correspond with numbers listed on installation diagram ADD(V)3299, provided with each kit.

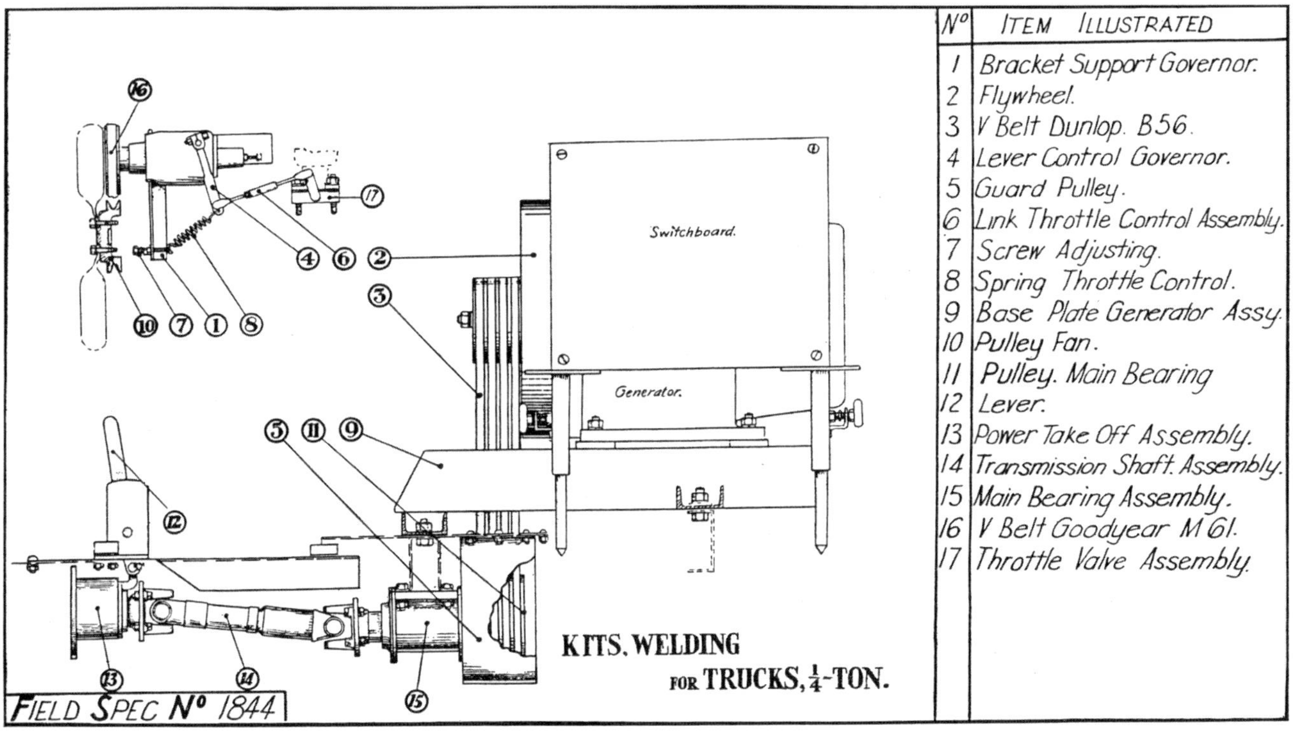

Notified in G.O.'s dated 30th November, 1944.

RESTRICTED.

The information given in this Document is not to be communicated, either directly or indirectly, to the Press or to any Person not authorized to receive it.

AUSTRALIAN MILITARY FORCES.

MECHANICAL VEHICLE FIELD SPECIFICATION NO. 1845.

KITS GENERATOR 5 K. V. A.

FOR FITMENT TO TRUCKS 1/4-TON 4 x 4

1944.

Prepared by Staff of the Master-General of the Ordnance and issued under the Direction of the Commander-in-Chief, Headquarters, Australian Military Forces.

By Authority: K. M. STEVENSON, Government Printer, Adelaide.

MECHANICAL VEHICLE FIELD SPECIFICATION.

General Details.

Army Nomenclature—Kits, Generator 5 KVA.

Role—For fitment to trucks, ¼ ton, 4 x 4.

Army Drawing No.—ADD (V) 3300.

Vocab Section for Kit—LV6-MT13.

Vocab No. for Kit—3819-Z.

EQUIPMENT, FITTINGS, AND STORES SUPPLIED WITH KIT.

Vocab Section.	Vocab No.	No. ID.	Designation.	Quantity.	See Note.*
—	—	1	Main bearing assembly	1	M
LV6-MT13	3888-Z	2	Power take-off assembly	1	M
LV6-MT13	3984-Z	3	Base plate generator	1	M
LV6-MT13	3879-Z	4	Levers operating PTO assembly	1	M
LV6-MT12	3915-Z	5	Valves, throttle assembly	1	M
LV6-MT13	3884-Z	6	Link, throttle control assembly	1	M
LV6-MT12	3907-Z	7	Governor assembly	1	M
LV6-MT13	3874-Z	8	Bracket, support governor	1	M
LV6-MT13	4007-Z	9	Bracket, mounting assembly, RH	2	M
LV6-MT13	4006-Z	10	Bracket, mounting assembly, LH	2	M
LV6-MT13	3985-Z	11	Pulley, generator	1	M
LV6-MT13	3987-Z	12	Stud, pulley	1	M
LV6-MT13	3986-Z	13	Pulley, main bearing	1	M
LV6-MT13	3878-Z	14	Guard, pulley	1	M
LV6-MT13	3903-Z	15	Transmission shaft assembly	1	M
—	N I V	16	Stud, carburettor	2	M
LV6-MT13	3885-Z	17	Spring, throttle control	1	M
—	N I V	18	Screw, adjusting	1	M
LV6-MT13	3901-Z	19	Pulley, fan	1	M
LV6-MT13	3875-Z	20	Cover, generator canvas, 5ft. 4in. square	1	M
—	N I V	21	Instruction plate	1	M
LV6-MT13	4007-Z	22	Bracket, switchboard, LH	1	M
LV6-MT13	4006-Z	22	Bracket, switchboard, RH	1	M
X3	XCA-1305A	23	Switchboard assembly	1	M
X3	XCA-1200 MA	24	Alternator assembly	1	M
LV6-MT1	3999-Z	25	Belt, " V " Dunlop, B52, or equivalent	3	M
LV6-MT1	3998-Z	26	Belt, " V " Goodyear, V. 10 or equivalent	1	M
G1	GAA6407	27	Screws, BSW (RD HD) M.S. ¼in. x 1½in.	4	M
G1	GA0700	28	Nuts, hexagonal, BSW, ¼in.	13	M
G1	GA1781	29	Washers, spring, single turn, ¼in diameter	15	M
G1	GAA5510	30	Bolts, BSW steel, hexagonal head, ⅜in. x 1in., with nuts	8	M
G1	GA1783	31	Washers, spring, single turn, ⅜in. diameter	28 (a)	M
—	N I V	32	Elbows, conduit, pressed metal, ¾in. diameter	2	M
G1	GAA5570	33	Bolts, BSW steel, hexagonal head, ⅜in. x 1¼in., with nuts	8	M
G1	GA0191	34	Bolts, BSW steel, hexagonal head, ½in. x 2¼in., with nuts	8 (e)	M
G1	GA1685	35	Washers, spring, single turn, ½in. diameter	13	M
G1	GAA5525	36	Bolts, BSW steel, hexagonal heads, ½in. x 1½in., with nuts	4 (e)	M
G1	GA0702	37	Nuts, hexagonal, BSW, ½in.	1	M
G1	GA1740	38	Washers, MS, ½in. diameter	1	M
G2	N I V	39	Key, steel, ⅜in. square, 1⅝in. long	1	M

A1845

Equipment, Fittings, and Stores Supplied with Kit—continued.

Vocab Section.	Vocab No.	No. ID.	Designation.	Quantity.	See Note.*
G1	GAA6405	40	Screws, BSW, round head, MS, $\frac{1}{4}$in. x $\frac{3}{4}$in. long	7	M
G1	GAA6410	41	Screws, BSW, round head, MS, $\frac{5}{16}$in. x 1in. long	6	M
G1	N I V	42	Bolts, SAE, hexagonal, MS, $\frac{3}{8}$in. x 1$\frac{1}{8}$in.	8 (b)	M
G1	GAA0283	43	Bolts, SAE, hexagonal, MS, $\frac{5}{16}$in. x 1in.	8 (c)	M
				(e)	M
G1	N I V	44	Screws, BSW, round head, MS, $\frac{5}{16}$in. x $\frac{3}{4}$in. long	7	M
G1	GAA6411	45	Screws, BSW, round head, MS, $\frac{5}{16}$in. x 1$\frac{1}{2}$in. long	2	M
G1	GA0699	46	Nuts, hexagonal, BSW, $\frac{5}{16}$in.	9	M
G1	GA1782	47	Washers, spring, single turn, $\frac{5}{16}$in. diameter ...	17 (d)	M
G1	GA0701	48	Nuts, SAE, hexagonal, $\frac{3}{8}$in.	2 (e)	M
G1	GAA6528	49	Screws, BSW, hexagonal, MS, $\frac{1}{4}$in. x 2in. long ..	4 (e)	M
G1	GAA6190	50	Screws, BA, round head, MS, No. 5, $\frac{1}{2}$in. long..	4 (e)	M
G1	GAA2732	51	Nuts, BA, hexagonal, MS, No. 5	4	M
G1	GAA4495	52	Washers, spring, single turn, $\frac{1}{8}$in. diameter	4	M

(a) 20 only required when ADD(V) 35-163 is used.
24 only required when ADD(V)35-164 is used.

(b) 4 only required when ADD(V)35-164 is used.
Not required when ADD(V)35-163 is used.

(c) Required only when ADD(V)35-163 is used.

(d) 25 required when ADD(V)35-163 is used.

(e) For items marked thus (e) maintenance replacements will be in accordance with Vocab Nos. as shown.

NOTE :—* This column denotes the source of initial supply of the items thus :—

O Supplied by Army Ordnance Service.

M Supplied by vehicle manufacturer or assembly contractor.

ID Numbers in this column correspond with numbers listed on installation diagram ADD(V)3300, provided with each kit.

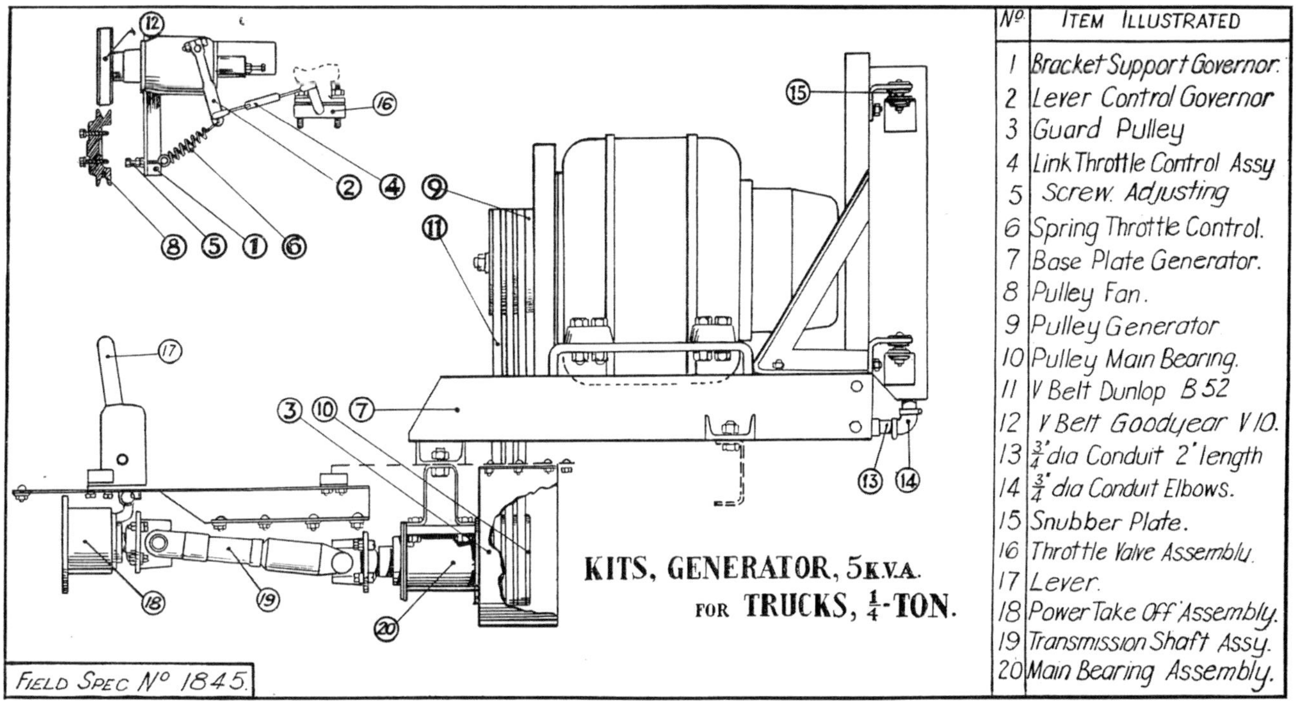

Notified in G.O.'s dated 31st December, 1944.

RESTRICTED.

The information given in this Document is not to be communicated, either directly or indirectly, to the Press or to any Person not authorised to receive it.

AUSTRALIAN MILITARY FORCES

MECHANICAL VEHICLE FIELD SPECIFICATION NO. 1846.

KITS GEAR SUPPORTING STRETCHERS LIGHT (AUST.) NO. 3.

FOR FITMENT TO TRUCKS 1/4-TON 4 x 4.

1944.

Prepared by Staff of the Master-General of the Ordnance and issued under the Direction of the Commander-in-Chief, Headquarters, Australian Military Forces.

Printed by Authority:
Sands & McDougall Pty. Ltd.

MECHANICAL VEHICLE FIELD SPECIFICATION

General Details

Army Nomenclature— Kits, Gear Supporting Stretcher Light (Aust.) No. 3.

Role— For fitment to trucks ¼ ton 4 x 4.

Army Drawing No.— ADD(V) 3318.

Vocab. Section for Kit— LV6-MT13.

Vocab. No. for Kit— 3752-Z.

EQUIPMENT FITTINGS AND STORES SUPPLIED WITH KIT

VOCAB. SECTION	VOCAB. Nos.	No.	ITEMS	QUANTITY	SEE NOTE *
LV6-MT13	3765-Z	1	Frame Assembly front	1	M
LV6-MT13	3774-Z	2	Frame Assembly rear	1	M
LV6-MT13	3775-Z	3	Runner Assembly	6	M
LV6-MT13	3762-Z	4	Clip frame front	6	M
LV6-MT13	3768-Z	5	Clip frame rear	4	M
LV6-MT13	3764-Z	6	Support Frame	5	M
LV6-MT13	3763-Z	7	Stiffener front	2	M
LV6-MT13	3769-Z	8	Stiffener rear	2	M
LV6-MT13	3770-Z	9	Stiffener rear centre upper	1	M
LV6-MT13	3771-Z	10	Stiffener rear centre lower	1	M
LV6-MT13	3772-Z	11	Stiffener support frame	2	M
LV6-MT13	3766-Z	12	Support spare wheel bracket near side ..	1	M
LV6-MT13	4207-Z	13	Support spare wheel bracket off side ..	1	M
LV6-MT13	3767-Z	14	"U" Bolt	2	M
LV6-MT13	4208-Z	15	Stay, near side	1	M
LV6-MT13	4209-Z	16	Stay, off side	1	M
G1	GAA5510	17	Bolts BSW steel hex head ⅜" x 1" ..	29	M
G1	GA1783	18	Washers spring single turn ⅜" dia. ..	25	M
G1	GA1738	19	Washers bolt ⅜" dia.	12	M
G1	GA0701	20	Nuts BSW Hex. ⅜"	4	M
G1	GA0160	21	Bolts BSW Hex. hd. 5/16" x 1" ..	12	M
G1	GA1782	22	Washers spring single turn 5/16" dia. ..	12	M
G1	GA0175	23	Bolts BSW Hex. hd. ⅜" x 3½"	4	M
G1	GAA5373	24	Bolts BSW Hex. hd. ½" x 1"	6	M
G1	GA1740	25	Washers bolt ½" diameter	6	M
G1	GA1785	26	Washers spring single turn ½" dia. ..	6	M

NOTE:— * This column denotes the source of initial supply:—

(M) Supplied by Vehicle Manufacturer or Assembly Contractor.

Numbers 1 to 26 correspond with numbers listed on Installation Diagram ADD(V)3318, provided with each kit.

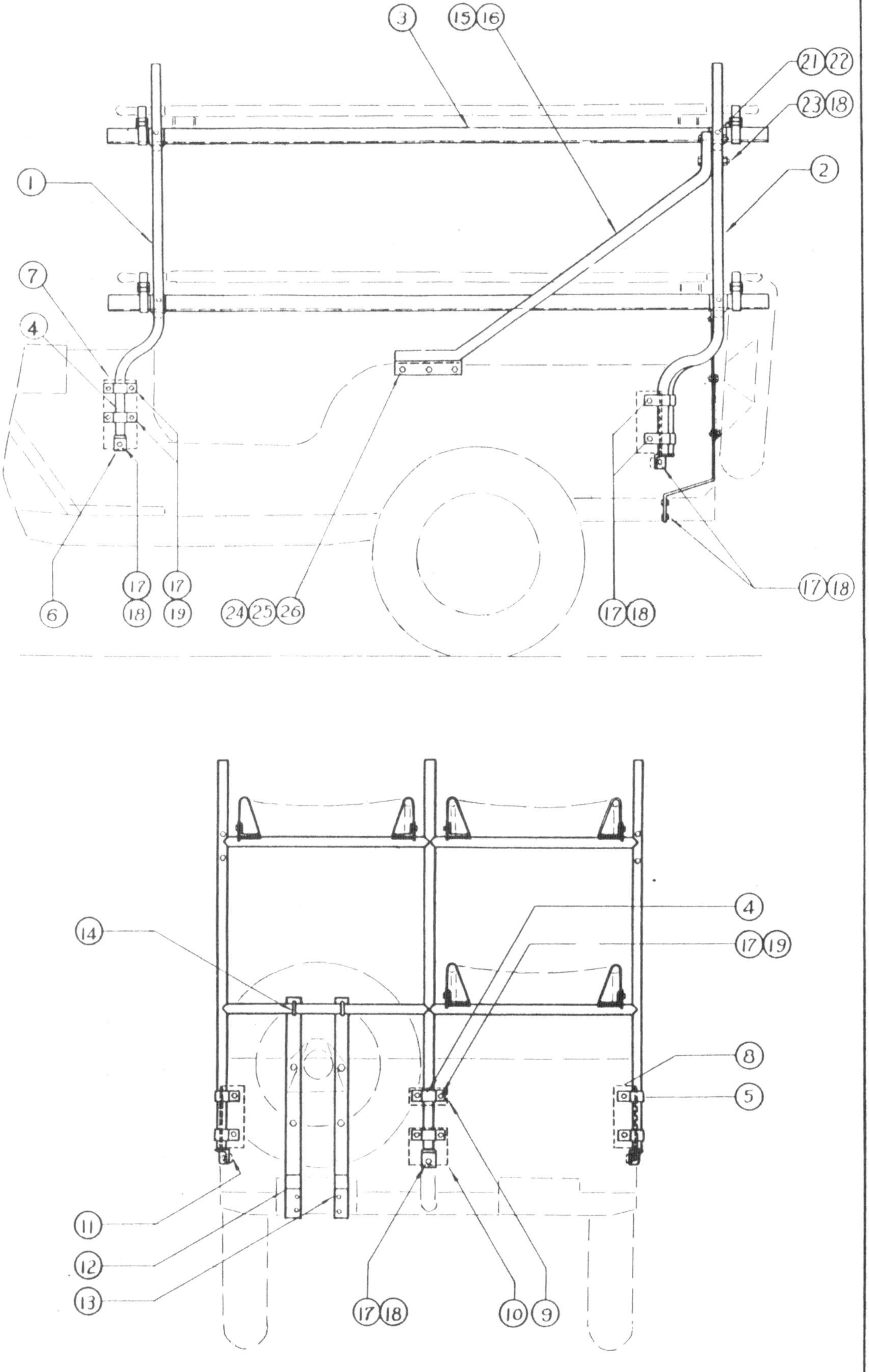

KITS – GEAR SUPPORTING STRETCHER
LIGHT AUST. Nº 3.

FIELD SPEC. Nº 1846

Notified in G.O.'s dated 30th November, 1944

RESTRICTED

The information given in this Document is not to be communicated, either directly or indirectly, to the Press or to any Person not authorised to receive it.

AUSTRALIAN MILITARY FORCES

MECHANICAL VEHICLE
FIELD SPECIFICATION NO. 1866.

KITS TANK 80-GALLONS - WATER

FOR FITMENT TO TRAILERS 2-WH. 8-CWT.

1944

Prepared by Staff of the Master-General of the Ordnance and issued under the Direction of the Commander-in-Chief, Headquarters, Australian Military Forces.

Lithographed by Authority
Modern Printing Co. Pty. Ltd.

MECHANICAL VEHICLE FIELD SPECIFICATION

General Details

Army Nomenclature - Kits, Tank 80-gals Water

Role - For fitment to trailers 2-wh 8-cwt

Army Drawing No. - ADD(V) 48-100

Vocab Section for Kit - LV6-MT13

Vocab No. for Kit - 4416-Z

Vocab Sect.	Vocab Nos	No.	Items	Quantity	See Note Ø
LV6-MT13	4427-Z	1	Tank assembly	1	M
LV6-MT13		2	Branch pipe assembly	1	M
LV6-MT13	4417-Z	3	Clamp pipe assembly	1	M
		4	Bases tank Hardwood 3'0½" x 5'11½" x 1⅜"	1(a)	M
LV6-MT13	4423-Z	5	Clips rope MS	4	M
		6	Packers clamp	1(b)	M
LV6-MT13	4424-Z	7	Rope ring sisal 6" O/D, 4½" I/D	4	M
		8	Foot valve strainer and float	1(c)	M
LV6-MT13	4425-Z	9	Hoses suction 1" x 21-ft long (rubber) with nut & tail & male end	1	M
G1	GA0016	10	Bolts BSW Steel,Cup head, sq. 5/16" x 3"	2	M
G1	GA1782	11	Washers spring single turn 5/16" dia.	26	M
G1	GA0038	12	Bolts BSW steel cuphead sq. 5/16" x 1¾"	24	M

NOTE: Ø This column denotes the source of initial supply:

(M) Supplied by vehicle manufacturer or assembly contractor

(a) In accordance with drawing ADD(V) 48-131

(b) In accordance with drawing ADD(V) 48-133

(c) In accordance with drawing MGO(M) 372/1.

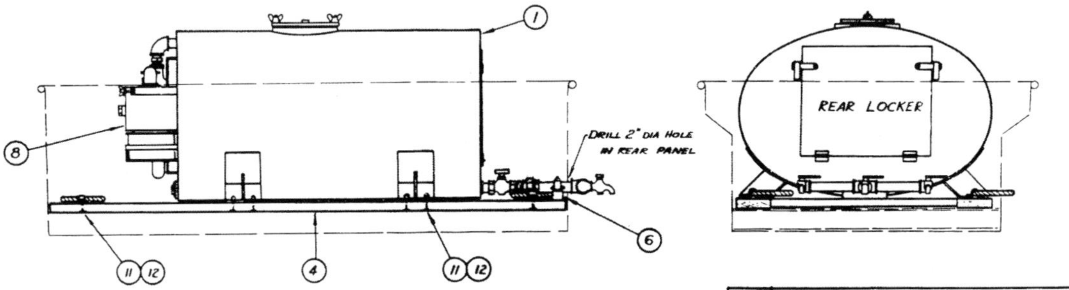

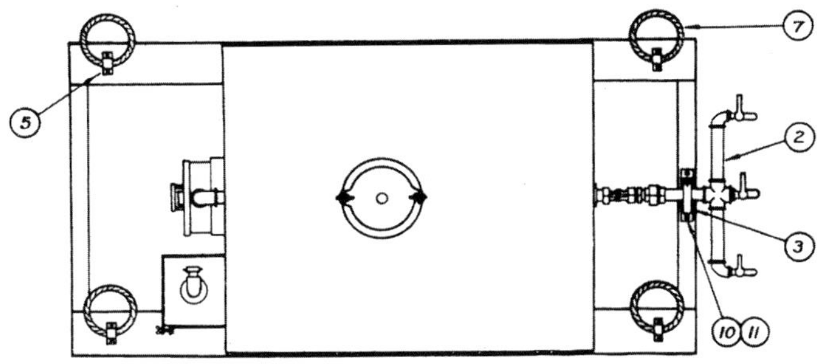

ITEM	DESCRIPTION	Nº REQD.
1	TANK ASSEMBLY.	1
2	BRANCH PIPE ASSEMBLY.	1
3	CLAMP, PIPE ASSEMBLY.	1
4	BASE.	1
5	CLIP ROPE.	4
6	PACKER.	1
7	ROPE RING.	4
8	FOOT VALVE, STRAINER & FLOAT.	1
9	21 FOOT LENGTH OF ¾" DIA. RUBBER SUCTION HOSE WITH NUT & TAIL & MALE END.	1
10	5/16" DIA. B.S.W. M.S. CGE. BOLTS × 2¾".	2
11	5/16" DIA. SPRING WASHERS.	26
12	5/16" DIA. B.S.W. M.S. CGE. BOLTS × 1¾", CAD. PL.	24

NOTE – ITEM 9 - SUCTION HOSE, & BRANCH PIPE TO BE HOUSED IN REAR LOCKER WHEN NOT IN USE.

FIELD SPEC. Nº 1866.

KITS TANK 80 GALS. WATER.

Notified in G.O.'s dated 31st December, 1944.

RESTRICTED.

The information given in this Document is not to be communicated, either directly or indirectly, to the Press or to any Person not authorised to receive it.

AUSTRALIAN MILITARY FORCES

MECHANICAL VEHICLE FIELD SPECIFICATION NO. 1885.

KITS MOBILE SERVICING.

1944.

Prepared by Staff of the Master-General of the Ordnance and issued under the Direction of the Commander-in-Chief, Headquarters, Australian Military Forces.

Printed by Authority:
Sands & McDougall Pty. Ltd.

MECHANICAL VEHICLE FIELD SPECIFICATION

General Details

Army Nomenclature— Kits, Mobile Servicing.

Role— Kit can be mounted on Trucks or Lorries G.S. Trucks ¼ ton or Trailers 2–wh. 8 cwt. and operates as a Mobile Unit used for servicing all types of Army Vehicles.

Army Drawing No.— ADD(V) 3321.

Vocab. Section for Kit— LV6–MT13.

Vocab. No. for Kit— 4136–Z.

EQUIPMENT, FITTINGS AND STORES SUPPLIED WITH KIT

VOCAB. SECTION	VOCAB. Nos.	No.	ITEMS	QUAN-TITY	SEE NOTE *
LV6–MT13	4140–Z	2	Frame base	1	M
LV6–MT13	4137–Z	3	Box Stores	1	M
LV6–MT13	4138–Z	4	Box, tools, stores and spares	1	M
		5	Pipe, Air delivery assembly	1	M
LV6–MT13	4139–Z	6	Bracket pipe air delivery	3	M
		7	Nipple galv. ½″ x ½″	1	M
		8	Elbow galv. ½″ x ½″	1	M
G1	GAA5510	9	Bolts BSW steel hex. hd. ⅜″ x 1″ with nuts	10	M
G1	GA0171	10	Bolts BSW steel hex. hd. ⅜″ x 1½″ with nuts	4	M
G1	GA1783	11	Washers spring single turn ⅜″ dia. ..	14	M
G1	GAA5614	12	Bolts BSW steel hex. hd. ¾″ x 1½″ with nuts	4	M
G1	GA1788	13	Washers spring single turn, ¾″ dia. ..	4	M
G1	GA0190	14	Bolts BSW steel hex. hd. ½″ x 2″ with nuts	6 (a)	M
G1	GA1785	15	Washers spring single turn ½″ dia. ..	6 (a)	M
G1	GA0195	16	Bolts BSW steel hex. hd. ½″ x 4½″ with nuts	6 (b)	M
G1	GA1740	17	Washers bolt ½″	6 (b)	M
LV6–MT6	33–T–234	18	Hose air line 3–ply ¼″ dia. 30 ft. long (lengths)	4	M
LV6–MT6	33–T–234	19	Hose air line 3–ply ¼″ dia. 1 ft. long (lengths)	8	M
LV6–MT1	0505–Z	20	Couplings, plug female bayonet type ¼″..	12	M
LV6–MT1	0507–Z	21	Couplings tails ¼″	16	M
LV6–MT1	0503–Z	22	Couplings body female bayonet type ¼″	8	M
LV6–MT1	0496–Z	23	Chucks air ¼″	2	M
LV6–MT1	0853–Z	24	Clamps Hose airline ¾″	24	M

(a) Not required when mounted on Trailers 2–wh. 8 cwt.

(b) Only required when mounted on Trailers 2–wh. 8 cwt.

(*) Denotes source of initial supply:—

(M) Supplied by Manufacturer or Assembly Contractor.

MVFS 1885 Kits Mobile Servicing

Page 1-

G1	GAA5614	Bolts, BSW, steel, hex hd, 3/4" x 1 1/2" with nuts	4 M
G1	GA 0195	Bolts, BSW, steel, hex hd, 1/2" x 4 1/2" with nuts	6(b)M
		Maintenance replacements for these two items to be-	
G1	GAA5530	Bolts, BSW, steel, hex hd, 3/4" x 2"	4 M
G1	GAA5595	Bolts, BSW, steel, hex hd, 1/2" x 5"	6(b)M

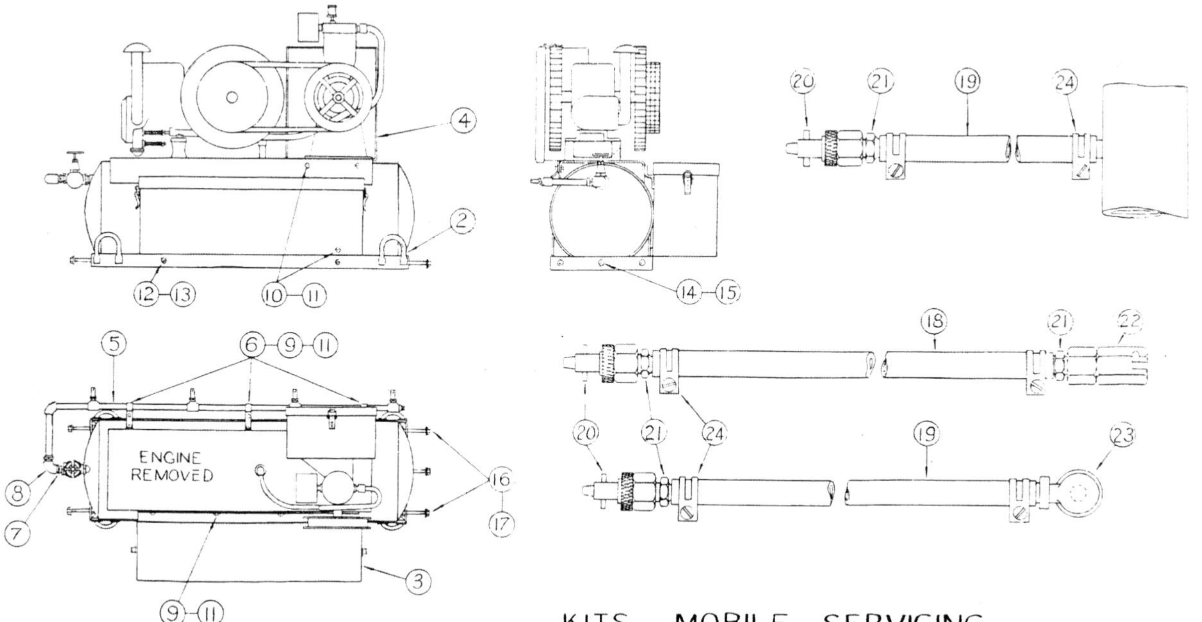

KITS MOBILE SERVICING

FIELD SPEC No 1885

APPENDIX 2:

Registration Quantities of Trucks 1/4 ton (USA)

Manufacturer	Army	RAAF	RAN	Total
Bantam Car Co	1	0	0	1
Willys-Overland Inc	6,747	379	51	7,177 (note 1)
Ford Motor Company	7,342	1,091	90	8,522 (note 2)
Maker not identified	32	10	4	46
Total	14,122	1,480	145	15,747 (note 3)

Note 1: Includes at least 3 Model MA.

Note 2: Includes 161 Model GPA

Note 3: This total is approximate and does not account for inter-service transfers, or for vehicles acquired overseas and not given an Australian registration number. Inter-service transfers are estimated to be around 500 vehicles. However, surviving records are not sufficiently detailed to allow an accurate correlation of chassis numbers between the three services (Army, RAAF, and RAN). Hence, the actual number of 'Trucks ¼ ton (USA)' received in Australia is estimated to be somewhere between 15,000 and 15,500 vehicles.

APPENDIX 3:

Registration of Trucks 1/4 ton Amphibious (Ford GPA)

Registration No. Range	Qty	Chassis No. Range	Comment
55701 to 55724	24		
58118 to 58143	26		
59327 to 59359	33		
126341 to 126346	6		
129944 to 129990	47		
130476 to 130500	25		130492, chassis number GPA9632, transferred to the Department of Air, and used at a flying boat base.
Total Army Registrations	161	GPA4749 to GPA10136	

APPENDIX 4:

Registration of Trailers, 8-cwt

Nomenclature/Type	**Manufacturer**	**Chassis Range**	**Qty Built/Recd**
Aust No.1	NSW L of C Area Wksp	Not numbered	20
Aust No.1	Peter Bros	4000 to 4069	70 *(note 1)*
Aust No.2 & 2A	Peter Bros	4000 to 4799; 9601 to 9670	800
Aust No.3 & 3A	Ford Motor Co	JT-1 to JT- 437	437
Aust No.3 & 3A	GMH	JT-1 to JT-671 (note 2)	672
Aust No.3 & 3A	Chas Hope	CH-1 to CH-200	200
Aust No.3 & 3A	IHC	IHC-1 to IHC-400	400
Aust No.3 & 3A	Sub-total		(1,709)
Aust No.4	Ford	JT-438 to JT-2087	1,654
Aust No.4	GMH	Same as Reg. Number	1,751
Aust No.4	Sub-total		(3,405)
Aust RAAF	Morse & Co	Not known	142
Aust No.2A for RAAF	Morse & Co	Not known	Not known: up to 50
TOTAL Aust Trailers	All Makes		6,196
TOTAL US Trailers	Bantam & Willys		4,461
TOTAL Not Known	Not Known *(note 3)*		214
TOTAL jeep Trailers			10,871

Notes:

(1) *The 870 figure may include 70 Aust No.1 trailers. From the surviving data, it is not possible to differentiate which of the jeep trailers built by Peter Bros were No.1 trailers, but as the company sequentially numbers all trailers, these are likely to be frame numbers 4000 to 4069.*

(2) *There is one trailer chassis number that is repeated in the GMH series, resulting in the last chassis number JT-671, being one less than the quantity supplied.*

(3) *The 'Not Known' group are those where the data that did not provide sufficient information to indicate either a country of origin or manufacturer. This means that figures quoted for each manufacturer should be taken as the minimum.*

APPENDIX 5:

Quantities of Trucks 1/4 ton and Trailers, 8-cwt in Australian Army service

Date	Trucks, 1/4 ton	Trailers, 8-cwt	Notes
1/6/1943	45%	35%	*See Note 1.*
1/8/1943	64%	35%	
1/9/1943	60%	na	
1/10/1943	64%	35%	
1/11/1943	70%	33%	Small increase in trailer allocations
1/12/1943	52%	22%	*See Note 2.*
1/2/1944	55%	40%	
8/7/1944	7,848 (104.4%)	4,358 (79.3%)	*See Note 3.*
19/8/1944	8,196 (113.1%)	5,282 (109.7%)	
2/9/1944	8,340 (116.6%)	5,516 (109.8%)	
7/10/1944	9,305 (128.8%)	6,062 (118.6%)	
14/10/1944	9,469 (130.8%)	6,038 (117.6%)	
21/10/1944	9,508 (132.9%)	5,221 (100.9%)	
28/10/1944	9,616 (134.5%)	5,225 100.9%)	
4/11/1944	9,632 (133.1%)	5,228 (100.2%	
18/11/1944	9,827 (135%)	6,076 (113.6%)	
25/11/1944	10,001 (137.4%)	6,084 (116.4%)	
2/12/1944	10,095 (132.4%)	5,128 (111.6%)	
9/12/1944	10,200 (133.8%)	5,247 (95.6%)	Increase in trailer entitlements
16/12/1944	10,280 (133.6%)	5,247 (95.6%)	Further allocations of trailers from USA production unlikely until 7/1945
30/12/1944	10,376 (134.8%)	5,247 (95.6%)	
6/1/1945	10,389 (135.3%	5,252 (94.9%)	
12/1/1945	10,389 (135.3%)	5,256 (95%)	
31/1/1945	10,662 (118%)	5,079 (72.3%)	Increase in trailer entitlements.
3/2/1945	10,646 (117.6%)	5,276 (69.9%)	
10/2/1945	10,926 (120.7%)	5,281 (69.9%)	
17/2/1945	11,113 (123%)	5,285 (69.9%)	
24/2/1945	11,130 (123.2%)	5,366 (71%)	
28/4/1945	11,532 (129%)	5,326 (70.8%)	
2/6/1945	11,468 (125.2%)	5,687 (74.2%)	
9/6/1945	11,522 (125.8%)	5,840 (76.2%)	
16/6/1945	11,581 (126.5%)	5,890 (76.8%)	
23/6/1945	11,653 (127.1%)	5,967 (76.7%)	
30/6/1945	11,780 (127.1%)	6,742 (86.7%)	
7/7/1945	11,736 (128.1%)	6,664 (85.7%)	
14/7/1945	11,868 (129.5%)	6,835 (87.9%)	

Date	Trucks, 1/4 ton	Trailers, 8-cwt	Notes
21/7/1945	12,016 (131.3%)	6,777 (87.2%)	
28/7/1945	12,102 (126.6%)	6,868 (88%)	
4/8/1945	12,111 (126.6%)	6,896 (88.4%)	
11/8/1945	12,133 (126.9%)	6,968 (89.2%)	
18/8/1945	12,156 (127.1%)	7,069 (90.4%)	15/8/1945: end of hostilities.
25/8/1945	12,141 (127%)	7,392 (93.9%)	

Notes:

(1) *Percentages are the quantity of jeeps and trailers available against the total quantity required to provide all units on the Order of Battle (OOB) with their full (100%) entitlements as listed in the War Equipment Table (WET) for each unit. There were constant adjustments to the full OOB entitlement as new units were raised, existing unit WETs were changed, or units were disbanded.*

(2) *Significant changes in the OOB increased the total entitlements. The jeep and trailer quantities available therefore constituted a much smaller percentage of the Army's full (100%) entitlement.*

(3) *Quantities are the number of jeeps and trailers in service at each date. The quantity is equal to the number of jeeps and trailers received into service minus losses from all causes, viz enemy action, accident, fair wear and tear, transfer to another service, or theft, which resulted in the jeep or trailer being struck off charge. The total quantity includes jeeps and trailers undergoing repair and not available for unit use.*

APPENDIX 6:

Examples of Trucks 1/4 ton (USA) used as RAN Ships Transport

Registration	Chassis No.	Ship HMAS	Notes
44058	GPW170001	Quickmatch	Former Army 146122
49057	nk	Warramunga	
51976	GPW114851	Barcoo	Former Army 58098, transferred 27/11/1946
51980	GPW175905	Quiberon	
51983	GPW213498	Australia	
52022	MB253227	Hobart	
52052	MB311964	Warramunga	
52058	nk	Tarakan	
52084	GPW167145	Australia	
52089	nk	Labuan	
52090	GPW254568	Sydney	Aboard during Korean war deployment
54219	MB123397	Bataan	
54241	GPW105650	Swan	

APPENDIX 7:

DME & RAAF Technical Instructions

BRAKES

COPY No. 2012

ALLIED LAND FORCES HEADQUARTERS

D.M.E. Technical Instruction

M.V. MODIFICATION

Instruction No. C.2–16 **A**

EQUIPMENT—TRUCKS, ¼-TON, 4 x 4. JEEP

SUBJECT—Hand brake—Cable adjustment.

REFERENCE—L.M. 1497.

---

1. *Purpose*—To provide increased brake cable adjustment.
2. *Drawings Required*—MGO(SK)1016 reproduced herein as Fig. 1.
3. *Items Affected*—
   - (a) New Parts—Nil.
   - (b) Replacement Parts—Nil.
   - (c) Substitute Parts—Nil.
   - (d) Discarded Parts—Nil.
   - (e) Supply of Parts—Nil.
   - (f) Material—4″ x ¾″ x ¼″ M.S. flat to be provided from local expense stocks.
4. *Special Tools, Jigs or Fixtures Required*—Nil.
5. *Instructions*—
   Manufacture plate and weld in position in accordance with Fig. 1.
6. *Remarks*—Log books will be endorsed on completion.

FILE UNDER—"BRAKES"

Frank Field

(FRANK FIELD) Brigadier,
Director of Mechanical Engineering.

Issue 1, 7th March, 1944 — Page 1.

By Authority: Victorian Railways Printing Works, North Melbourne. 305/44.

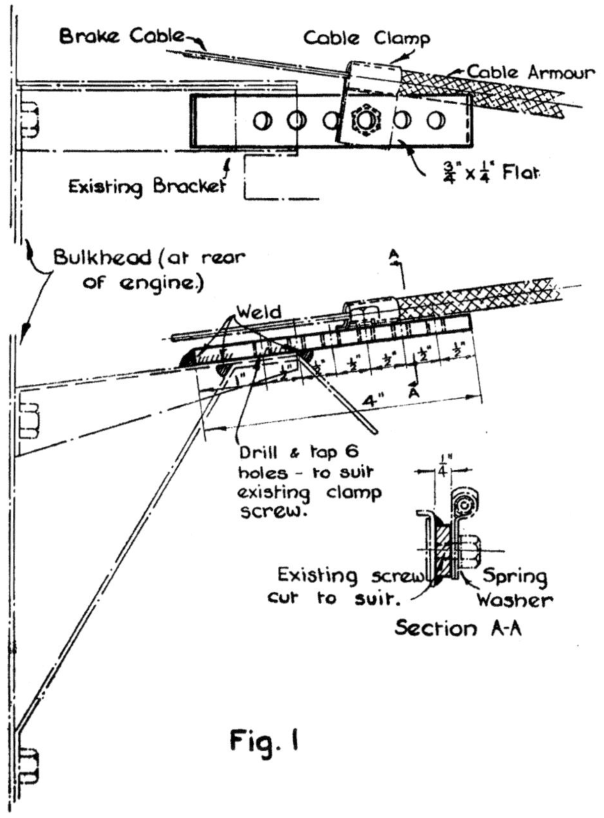

Fig. 1

Issue 1, 7th March, 1944

FRONT AXLE.

ALLIED LAND FORCES HEADQUARTERS

Instruction No.

D.M.E. Technical Instruction

C3-8

B

B

MV MODIFICATION

FILE UNDER - "FRONT AXLE"

EQUIPMENT:— TRAILERS, 2-WHEELED, 8-CWT., (AUST) NO. 1.

SUBJECT:— Conversion to Trailers, 2-wheeled, 8-cwt., (Aust) No. 2.

REFERENCE:— L.M.1684

1. *Purpose:-* Conversion of trailer.

2. *Drawings required:-* Fig. 1 reproduced herein.

3. *Items affected:-*
   (a) New Parts:- Nil.
   (b) Replacement Parts:-

Vocab No.	Designation	Qty. per Equipt.
	SECTION LV7	
	Trailer, 2-wheeled, 8-cwt., (Aust) No.2	
2801-Z	Kit, for conversion of Trailer, 2-wheeled, 8-cwt., (Aust) No. 1 to (Aust.) No. 2	1
	Consisting of -	
1789-Z	Axle	1
1934-Z	Bearing, wheel, inner complete (909002ND)	2
1933-Z	" " outer complete (909001ND)	2
	Bolts, ANF., 1/2" x 5-1/2" hex. hd. with nuts	8
2504-Z	Bolts, shackle 9/16" dia. x 3-1/8" long	6
2507-Z	Bush, spring eye	4
405999	Caps, hub	2
1790-Z	Hubs, wheel	2
45-F-400	Nipples, lubricating, hydraulic Zerk straight	6
1830-Z	Nuts, axle, special	2
	Nuts, castle, 9/16" ANF	6
405984	Nuts, disc wheel	10
	Nuts, lock, ANF. 1/2" hex.	8
	Pins, keep, split 1/8" x 1-1/2"	6
	" " " 1/8" x 1-3/4"	2
1809-Z	Plate, clamp, spring	2
1808-Z	" " "	2
2506-Z	Plate, shackle, inside	2
2505-Z	" " outside with stop	2
M6007	Retainer, grease, wheel hub	2

Vocab No.	Designation	Qty. per Equipt.
	SECTION LV7 (Contd.)	
	Trailer, 2-wheeled, 8-cwt., (Aust) No. 2, (Contd.)	
	Kit, for conversion of Trailer, 2-wheeled, 8-cwt., (Aust) No. 1 to (Aust.) No. 2 (Contd.)	
	Consisting of - (Contd.)	
2503-Z	Shackles, front	2
2502-Z	Shackles, rear, with bush	2
1792-Z	Spring, assembly and plates	2
499542	Studs, std., knurled	10
1932	Washers, tongue, std., 3/4" dia.	2
1787-Z	Wheels, 6.00 x 16	3

(c) Substitute Parts:- Nil.
(d) Discarded Parts:- To be forwarded to Ordnance Depots to be held for issue as spare parts.
(e) Supply of Parts:- Kits will be supplied under LHQ arrangements to Ordnance Depots whence they will be demanded by Officers concerned.
(f) Material:- Nil.

4. *Special Tools, Jigs or Fixtures required:-* Nil

5. *Instructions:-* Remove existing wheels, axles, hubs and springs and replace with new assembly as shown in Fig. 1.

6. *Remarks:-* Nil.

Frank Field

(FRANK FIELD) Brigadier
Director of Mechanical Engineering

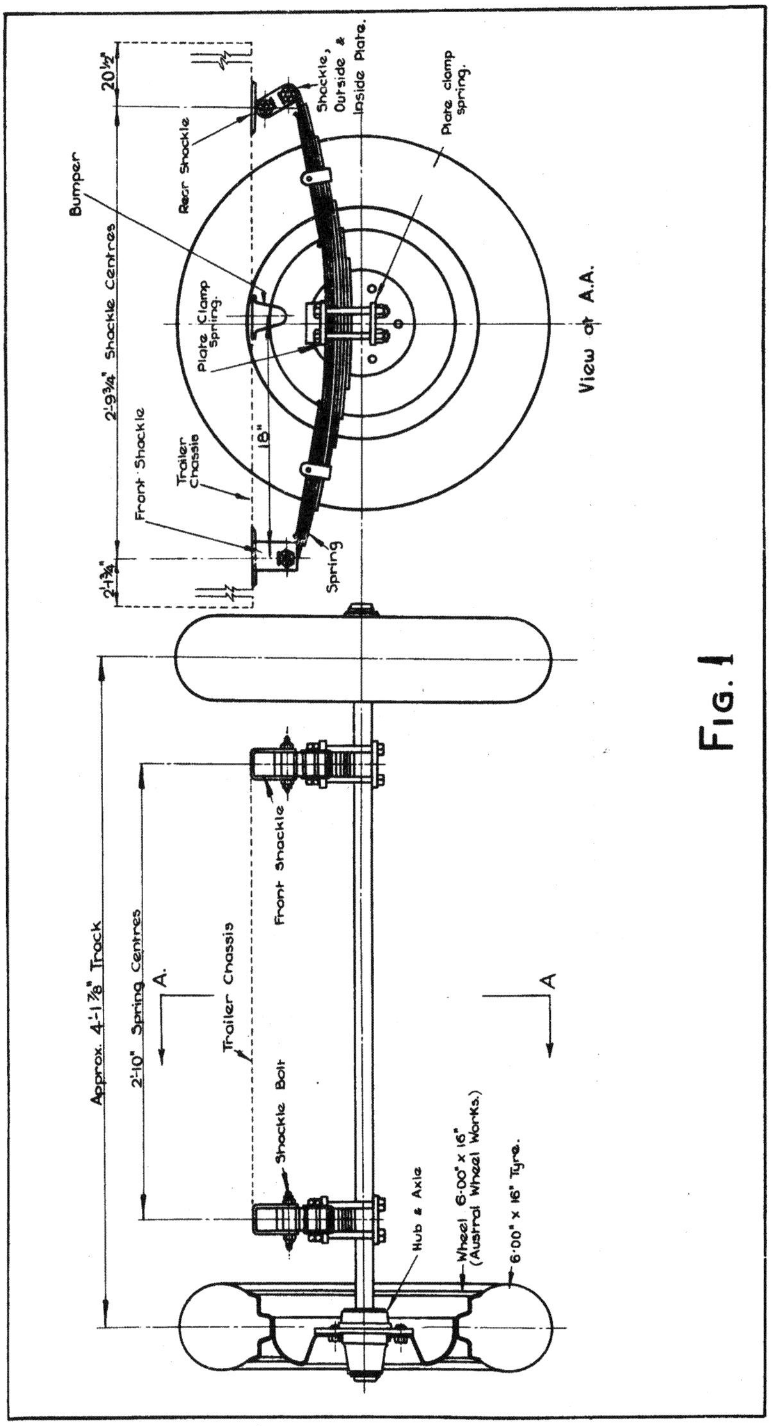
20½"
Shackle, Outside & Inside Plate.
Plate clamp spring.
Rear Shackle
Bumper
2'-9¾" Shackle Centres
Plate Clamp Spring.
View at A.A.
18"
Trailer Chassis
Front Shackle
2'-1¾"
Spring
Approx. 4'-1⅞" Track
2'-10" Spring Centres
A.
Trailer Chassis
Front Shackle
Shackle Bolt
A
Hub & Axle
Wheel 6·00" x 16" (Austral Wheel Works.)
6·00" x 16" Tyre.

Fig. 1

COPY No 2287

ALLIED LAND FORCES HEADQUARTERS

D.M.E. Technical Instruction

Instruction No. C.5–12 **B**

Workshop Issue **B**

M.V. REPAIR

EQUIPMENT—TRUCKS, ¼-TON.

SUBJECT—Strengthening of drag link bell crank.

REFERENCE—L.M. 1519.

1. *Purpose*—To prevent fracture of either arm of the bell crank.
2. *Drawings Required*—MGO(SK)1032 reproduced herein.
3. *Items Affected*—
   - (a) New Parts—Nil.
   - (b) Replacement Parts—Nil.
   - (c) Substitute Parts—Nil.
   - (d) Discarded Parts—Nil.
   - (e) Supply of Parts—Nil. Existing components to be modified.
   - (f) Material—3/16″ M.S. plate to be provided from local expense stocks.
4. *Special Tools, Jigs or Fixtures Required*—Nil.
5. *Instructions*—
   Weld 3/16″ M.S. gusset plate between the two arms of the bell crank as shown in drawing.
6. *Remarks*—Nil.

FILE UNDER—"STEERING"

Brigadier,
Director of Mechanical Engineering

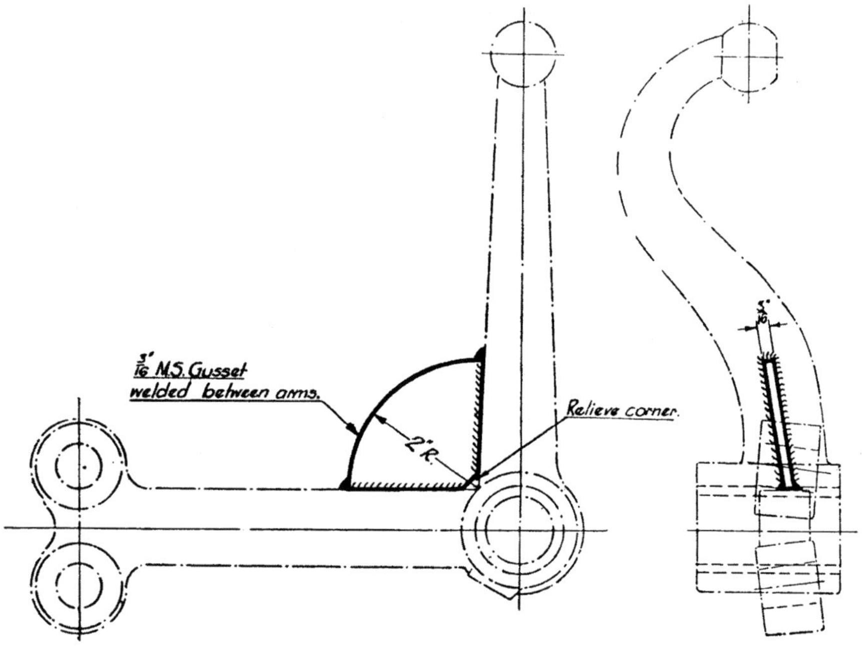

TRUCKS ¼ TON.
Modification to Drag Link Bell Crank.

M.G.O.(SK) 1032.

By Authority: Victorian Railways Printing Works, North Melbourne. 179/44.

ALLIED LAND FORCES HEADQUARTERS

Instruction No. C.6–8

D D

D.M.E. Technical Instruction

M.V. MODIFICATION

EQUIPMENT—TRAILERS, 2-WHEELED, WATER, 80 GALLON (AUST.).

SUBJECT—Addition of strengthening gussets.

REFERENCE—L.M. 1409.

1. *Purpose*—To prevent failure of main frame.
2. *Drawings Required*—MGO(V)547C reproduced herein as Fig. 1.
3. *Items Affected*—
   (a) New Parts—Nil.
   (b) Replacement Parts—Nil.
   (c) Substitute Parts—Nil.
   (d) Discarded Parts—Nil.
   (e) Supply of Parts—Nil.
   (f) Material—To be obtained from local expense stocks.
4. *Special Tools, Jigs or Fixtures Required*—Nil.
5. *Instructions*—Weld 2″ x 5/16″ x 2′6″ stiffener plates to top and bottom of each side main frame members, as shown in Fig. 1.
6. *Remarks*—Nil.

FILE UNDER—"FRAME."

Frank Field

(FRANK FIELD) Brigadier,
Director of Mechanical Engineering.

Issue 1, 24th August, 1944.

By Authority: Victorian Railways Printing Works, North Melbourne. 3151/44.

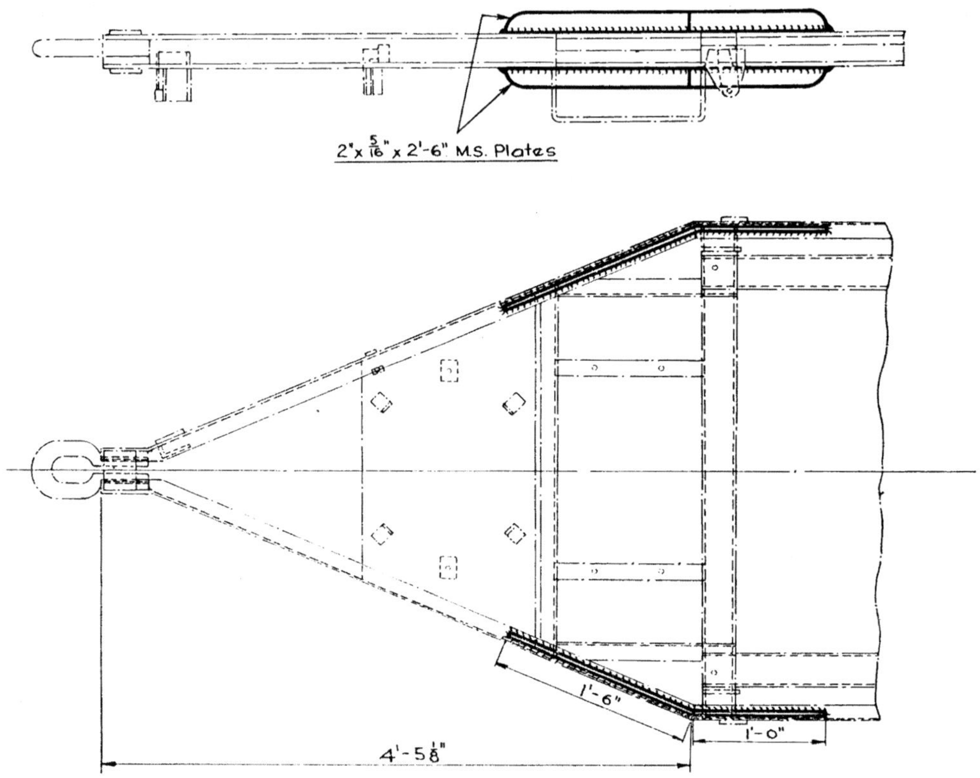

FIG. 1.

EQUIPMENT—TRUCKS, ¼-TON.

SUBJECT—Frame assembly—Correction to dimensions.

REFERENCE—Army Motors Vol. 4, Sept., 1943.

1. The dimensions shown in the figure appearing on page 1500-120 of Maintenance Manual of Willys Truck are incorrect.

2. The correct dimensions are tabulated in Fig. 1 reproduced herein.

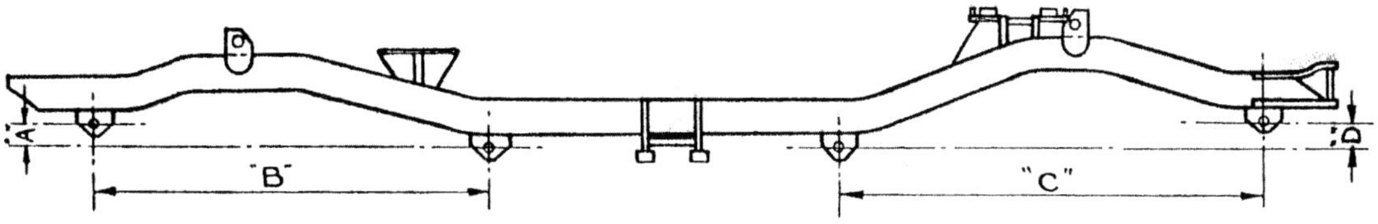

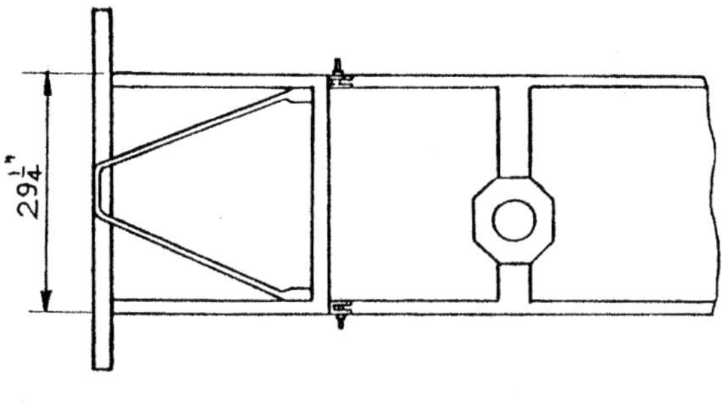

"A"	4"	should be changed to	3 3/8"
"B"	34 7/8"	" " " "	40 1/16"
"C"	40 1/16"	" " " "	34 3/4"
"D"	3 3/8"	" " " "	4 5/32"

FIG. I.

COPY N° 2126

SPRINGS

ALLIED LAND FORCES HEADQUARTERS **Instruction No.**

D.M.E. Technical Instruction

C.7–10

Workshop Issue

M.V. MAINTENANCE

EQUIPMENT—TRUCKS, WILLYS, $\frac{1}{4}$-TON 4 × 4.

SUBJECT—Shock Absorber Adjustment.

REFERENCE—Maintenance Manual TM10-1207.

Failures of the shock absorbers on the above vehicles are excessive, and from an examination of those that have failed, it appears that the major cause of failure is due to using excessive force when adjusting them.

2. The shock absorbers are a direct action type, two-way control, and adjustable, and provide a much smoother ride by dampening the spring action as the vehicle passes over irregularities in the road.

3. *Adjustment of shock absorber,* Refer to Figs. 1 and 2.

(a) Remove the lower end from the spring plate, push the unit together to engage the adjusting key, and turn in a clockwise direction until the limit of adjustment is reached.

(b) Holding the adjusting key in slot, turn the lower end anti-clockwise two turns. This is the average adjustment.

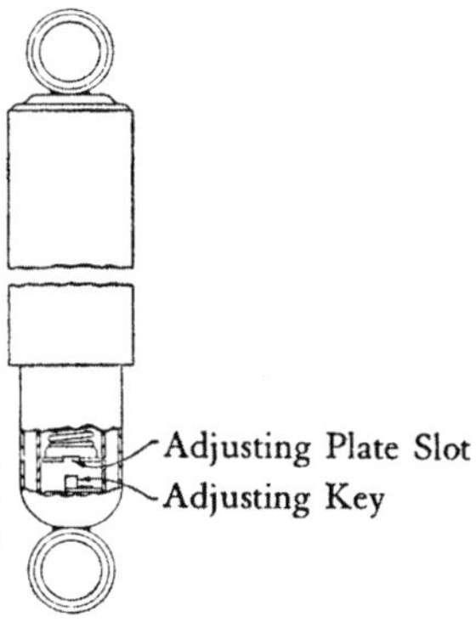

Fig. 1

Sketch showing shock absorber before engaging the adjusting slot and key.

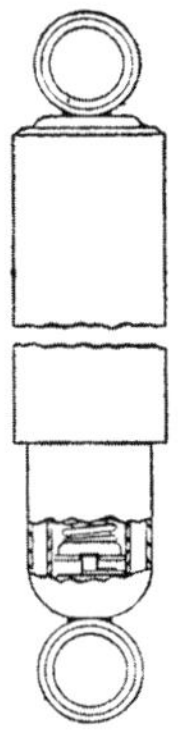

Fig. 2

Sketch showing shock absorber completely collapsed with adjusting key engaged in adjusting plate slot.

FILE UNDER—"SPRINGS"

Note.—The adjusting key shown is secured by a small rivet, and if *excessive pressure* is used in the adjustment, this key is pulled from the plate. Attached to the key is a flat spring valve which, when removed, renders the shock absorber unserviceable.

4. Turning the adjustment to the right or clockwise gives a firmer control for rough roads, while turning in the opposite direction gives a softer control, allowing faster spring action.

5. Should squeaks occur in the rubber mounting bushing, do not use mineral oil or rubber lubricant, but add a flat washer on the mounting pins to place the bushing under pressure, and prevent movement between the rubber and metal part.

Brigadier,
Director of Mechanical Engineering

Issue 1, 29th January, 1944.

Page 1

EQUIPMENT—TRUCK, WILLYS, ¼-TON 4 × 4, MODEL MB.

SUBJECT—Flywheel and Crankshaft Flange.

REFERENCE—D.D.M.E. Tech. Inst. No. NA/TI/173.

In production, the flywheel and crankshaft flange are now each marked with an arrow, to facilitate assembly. In reassembling these parts, care should be taken to ensure that the two arrows are assembled together.

ENGINE

ALLIED LAND FORCES HEADQUARTERS

Instruction No.

D.M.E. Technical Instruction

C8-69

MV MAINTENANCE

WORKSHOP ISSUE

EQUIPMENT:— TRUCKS, 1/4-TON.

SUBJECT:— Engine Sump Drain Plug Shield

REFERENCE:—

FILE UNDER - "ENGINE"

Damage to the sump plug boss and the loss of drain plugs has been experienced due to inadequate protection provided by the sump protection plate.

In some cases, the sump drain plug is located below the protection plate, and either becomes loose due to road contact, or the plug boss is severely damaged.

Where conditions may warrant added protection, the fitting of a shield similar to that shown in Fig. 1 should eliminate further trouble.

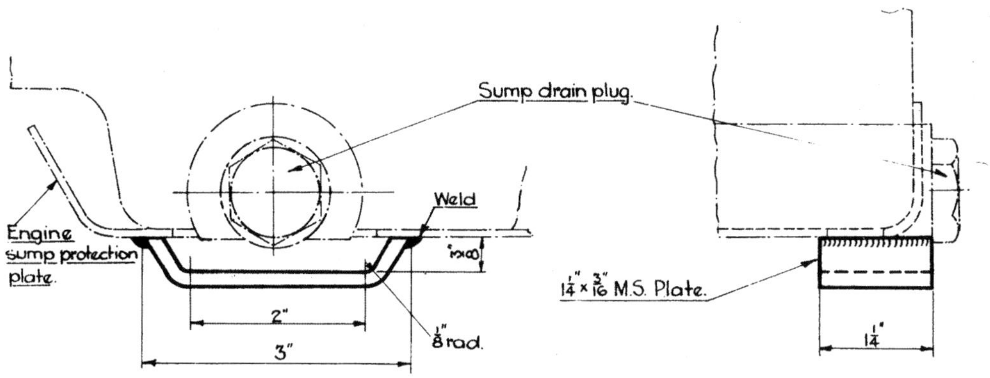

FIG. I

(FRANK FIELD) Brigadier,
Director of Mechanical Engineering.

Issue 1, 16th May 1944.

L.H.Q. MISC 6457 h

AUSTRALIAN MILITARY FORCES
Master-General of the Ordnance

Instruction No.
C8-74

D.M.E. Technical Instruction

M.V. REPAIR

B

EQUIPMENT:- TRUCKS, 1/4-TON, 4 x 4, JEEP - FORD AND WILLYS

SUBJECT:- Exhaust manifold heat control valve.

REFERENCE:-

1. Purpose:- To lock the exhaust manifold heat control valve open.

2. Instructions:-

(a) With the exhaust heat control valve in the open position, tack weld the valve spindle to the manifold boss, using bronze rod.

(b) Care will be taken to ensure that excessive heat is not applied to the manifold proper or cracking may occur.

3. Remarks:- Operating under humid conditions, it has been found that the valve mentioned rusts and seizes in the closed position. As the valve is not considered necessary in semi-tropical or tropical areas, workshops will carry out the above instructions when engines are being overhauled and ONLY where the vehicles are used or are to be used north of 20° south latitude or tropical areas outside Australia.

FILE UNDER - "ENGINES"

M Short. Lt. Col. for Brigadier,
(C.A. GEYTON)
Director of Mechanical Engineering.

TRANSMISSION & CLUTCH

COPY **Nº** 2484

ALLIED LAND FORCES HEADQUARTERS

D.M.E. Technical Instruction

Instruction No. C.9–37

B B B B

M.V. MODIFICATION

EQUIPMENT—TRUCKS, ¼-TON.

SUBJECT—Oil Seal Protectors—Fitting to front and rear Universal Joints.

REFERENCE—L.M. 1648.

1. *Purpose*—To prevent damage to oil seals.
2. *Drawings Required*—Fig. 1 reproduced herein.
3. *Items Affected*—
   (a) New Parts—

Part No.	Designation.	Qty. per Equipt.
3545-Z	Protectors, oil seal	2

   (b) Replacement Parts—Nil.
   (c) Substitute Parts—Nil.
   (d) Discarded Parts—Nil.
   (e) Supply of Parts—Parts listed under para. 3(a) will be demanded from Ordnance Depots by A.E.M.E. Officers on the scale shown.
   (f) Material—Nil.
4. *Special Tools, Jigs or Fixtures Required*—Nil.
5. *Instructions*—Dismantle universal joints, fit protectors as shown in Fig. 1, and reassemble.
6. *Remarks—Nil.*

FILE UNDER—"TRANSMISSION AND CLUTCH"

(FRANK FIELD) Brigadier,
Director of Mechanical Engineering.

Issue 1, 19th May, 1944 Page 1.

By Authority: Victorian Railways Printing Works, North Melbourne. 2086/44.

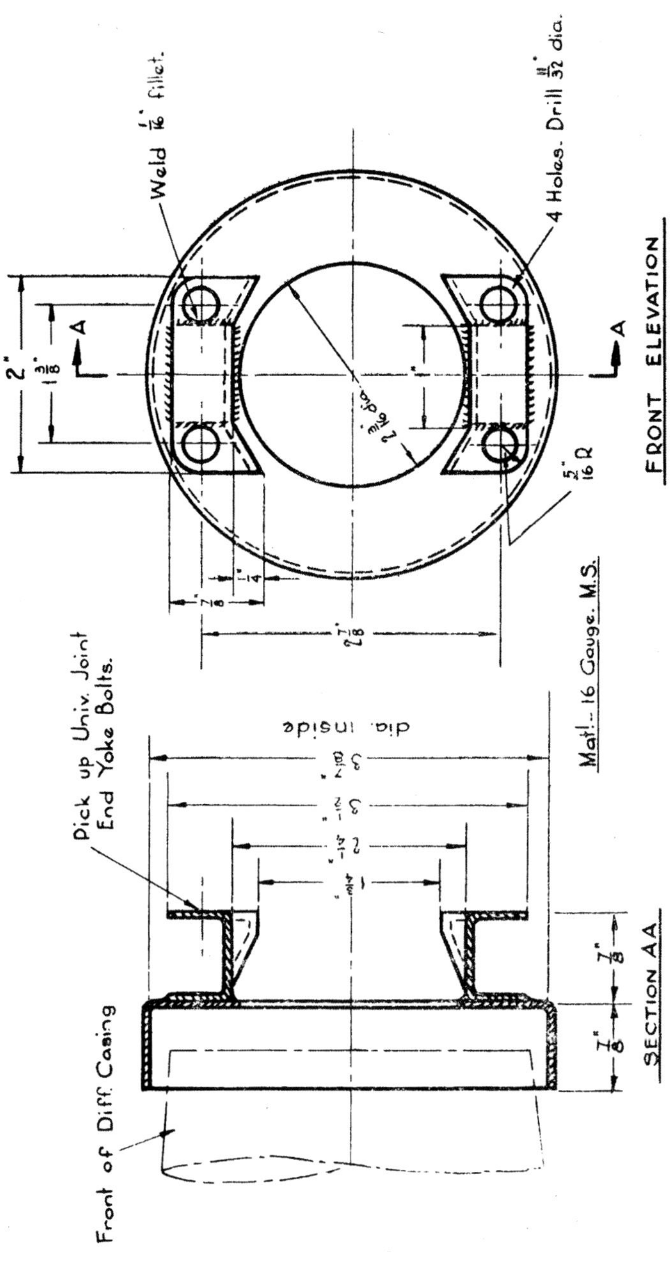

TRUCKS 1/4 TON 4x4 (U.S.A.)
DIFFERENTIAL OIL SEAL PROTECTOR.

FIG. 1.

ALLIED LAND FORCES HEADQUARTERS

Instruction No. C.9–46

C

C

D.M.E. Technical Instruction

M.V. MODIFICATION

EQUIPMENT—TRUCKS, ¼-TON (FORD AND WILLYS).

SUBJECT—Countershaft Gear and Thrust Washers (Warner Gear Box).

REFERENCE—L.M. 1909.

1. *Purpose*—To reduce wear on countershaft and gearbox thrust faces, by providing improved lubrication.
2. *Drawings Required*—Fig. 1 reproduced herein.
3. *Items Affected*—
   (a) New Parts—Nil.
   (b) Replacement Parts—Nil.
   (c) Substitute Parts—Nil.
   (d) Discarded Parts—Nil.
   (e) Supply of Parts—Nil. Existing components to be modified.
   (f) Material—Nil.
4. *Special Tools, Jigs or Fixtures Required*—Nil.
5. *Instructions*—
   (a) Oil grooves will be cut on the outside of both countershaft gear bushings (Ford Part No. GWP-7121, Willys Part No. A-878) as shown in (c) Fig. 1.
   (b) Oil scoop grooves will be ground on both faces of the transmission counter gear, in the direction of rotation as indicated in (b) Fig. 1.
   (c) Oil scoop grooves will be cut on the rear face of the transmission countershaft thrust washer —Rear (Bronze) (Ford Part No. GWP-7129, Willys Part No. 635811) in the direction of rotation indicated in (a) Fig. 1.
6. *Remarks*—Nil.

FILE UNDER—"TRANSMISSION & CLUTCH"

(FRANK FIELD) Brigadier,
Director of Mechanical Engineering.

By Authority: Victorian Railways Printing Works, North Melbourne. 3152/44.

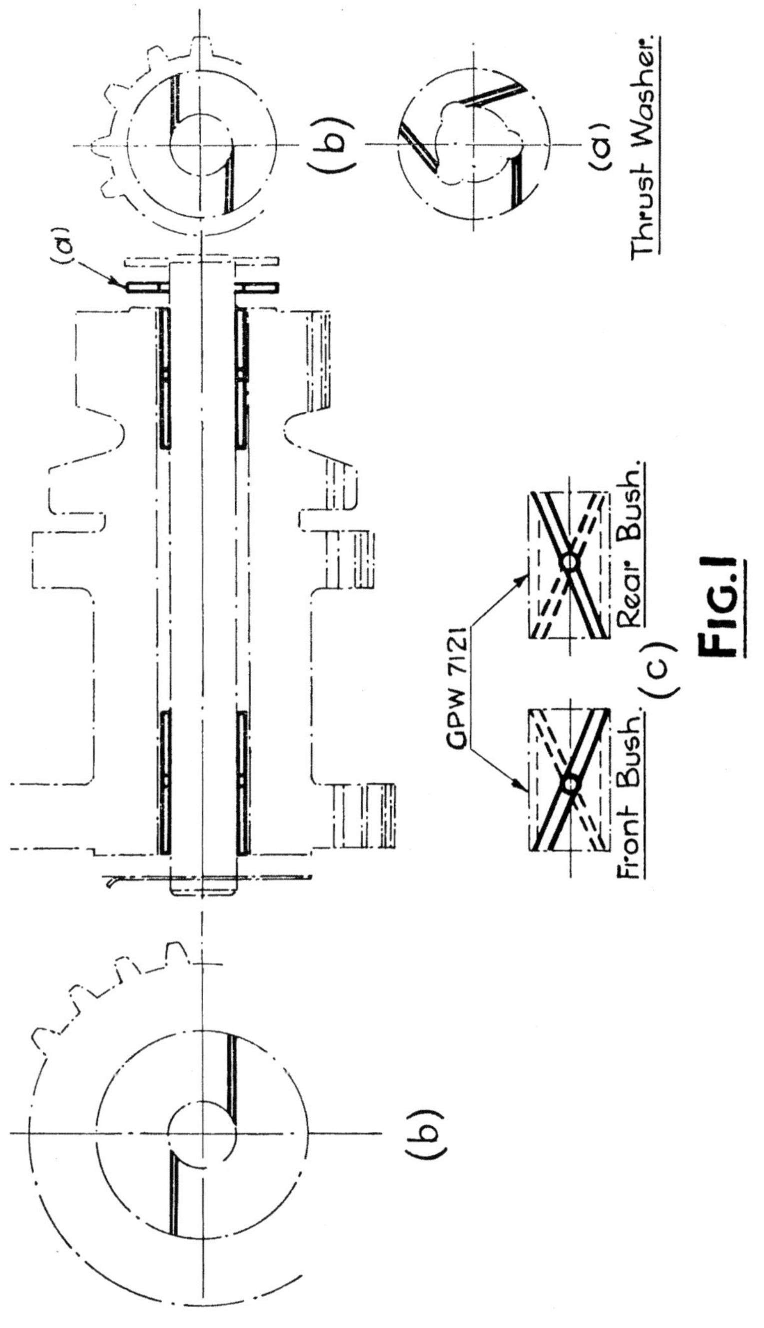

FIG.1

ME6u/BH

ALLIED LAND FORCES HEADQUARTERS — Instruction No. C11-21 [A]

D.M.E. TECHNICAL INSTRUCTION

M.V. MODIFICATION

EQUIPMENT:- TRUCKS, 1/4-TON.

SUBJECT:- Fuel pump - Modification.

REFERENCE:- L.M. 1464.

---

1. Purpose:- To prevent leakage of petrol past the fuel pump diaphragm on 1/4-ton Trucks (Jeeps).

2. Drawings Required:- Fig.1 reproduced herein.

3. Items Affected:-

   (a) New Parts:- Nil
   (b) Replacement Parts:- Nil
   (c) Substitute Parts:- Nil
   (d) Discarded Parts:- Nil
   (e) Supply of Parts:- Nil - Existing components to be modified.
   (f) Material:- Nil

4. Special Tools, Jigs or Fixtures Required:- Nil

5. Instructions:-

   Drill a 5/32" drain hole at "B" through into recess "A" as shown in Fig.1.

6. Remarks:-

   In some castings a hole is already in existence, but does not continue through to the recess. In this case the hole will need to be drilled approximately 3/32" deeper, using a 5/32" drill.

A.R. McEwen Col

for (FRANK FIELD) Brigadier,
Director of Mechanical Engineering.

27/9/43.

(P.T.O.)

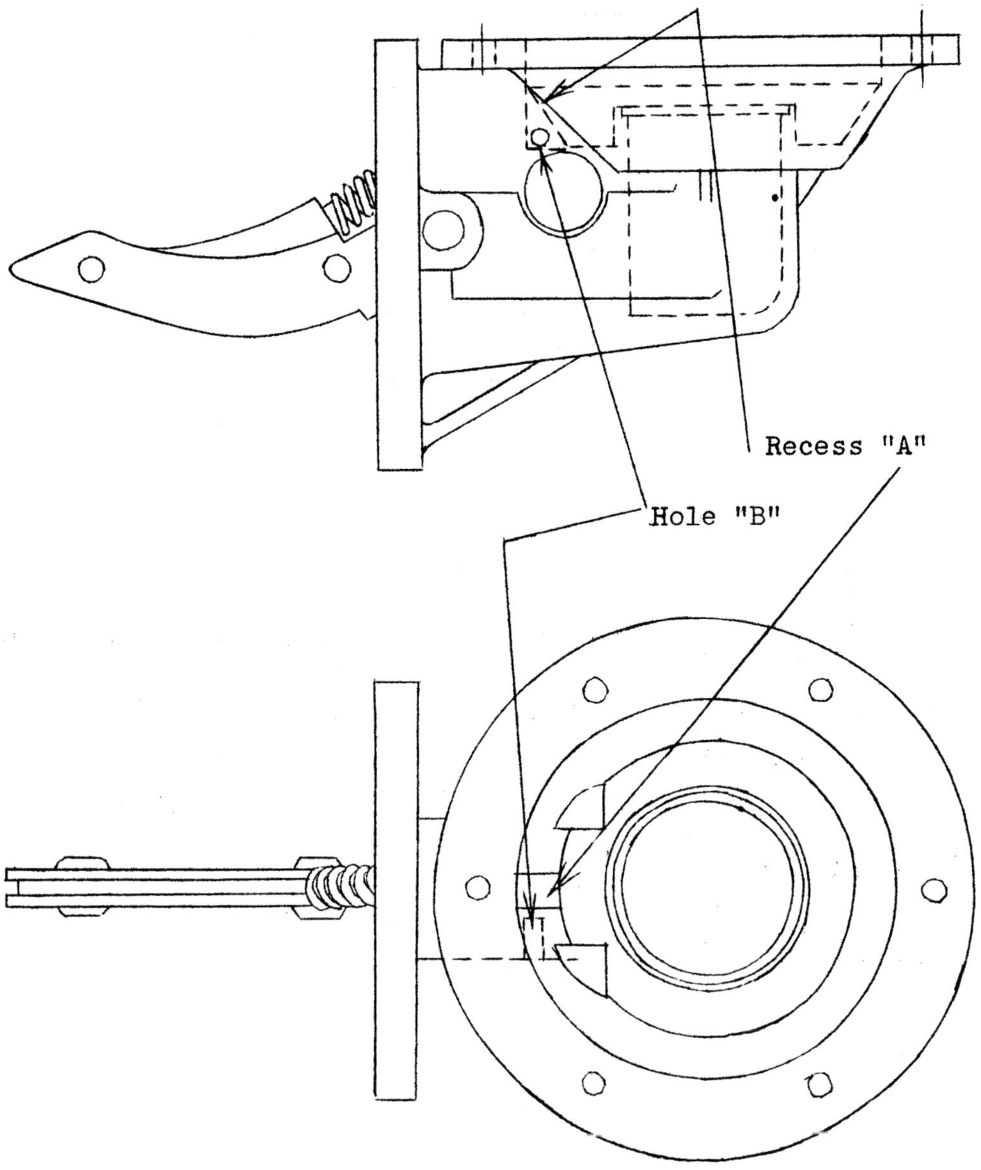

FIG. 1

FUEL PUMP - TRUCK 1/4-TON.

Copy No. ....

ELECTRICAL SYSTEM

Instruction No.

C12-38

ALLIED LAND FORCES HEADQUARTERS

DME TECHNICAL INSTRUCTION

MV MODIFICATION

EQUIPMENT:- TRAILERS, 2-WHEELED, 8-CWT (USA)

SUBJECT:- Tail lamps

REFERENCE:- E.M. 1253

---

1. Purpose:- To remove tail lamps.

2. Drawings Required:- Nil.

3. Items Affected:-

(a) New Parts:- Nil.
(b) Replacement Parts:- Nil.
(c) Substitute Parts:- Nil.
(d) Discarded Parts:-
  Wiring system comprising -
    Blackout switch
    Coupling socket plug and cable assembly
    Connector (3 way)
    Cable, right
    Cable, left
    Tail light assembly, right
    Tail light assembly, left
  To be forwarded to RSD
(e) Supply of Parts:- Nil.
(f) Material:- Nil.

4. Special Tools, Jigs or Fixtures Required:- Nil.

5. Instructions:- Remove parts to be discarded and pack in suitable containers for transit.

6. Remarks:- Nil.

Brigadier,
(FRANK FIELD)
Director of Mechanical Engineering.

File under:- ELECTRICAL SYSTEM

Issue 1. 11th November, 1944. Page 1.

AUSTRALIAN MILITARY FORCES
Master-General of the Ordnance

Instruction No.

D.M.E. Technical Instruction C12-39

M.V. INFORMATION

EQUIPMENT:- TRUCKS, 1/4-TON, FORD AND WILLYS

SUBJECT:- Voltage regulators - Repair kits.

REFERENCE:-

FILE UNDER - "ELECTRICAL"

1. Kits will shortly be available for the repair of Jeep voltage regulators comprising items as listed hereunder. Kits only are supplied and indents will, under no circumstances, be placed for individual parts.

Part No.	Designation
A7085	Kit, repair, current regulator - Comprising - 1 A7808 Armature 1 A7796 Nut, adjusting 1 A7797 Screw, adjusting 2 A7802 Screw 1 A7806 Spring 3 A1666 Washer 3 51532 Washer, lock
A7794	Kit, repair, generator cut-out relay - Comprising - 1 A7800 Armature 1 A7801 Bracket and contact 1 A7799 Insulation 1 A7796 Nut, adjusting 1 A7797 Screw, adjusting 2 A7802 Screw 2 A7803 Screw 1 A7798 Spring, armature 3 A1666 Washer 2 A7795 Washer 2 A7804 Washer 3 51532 Washer
A7810	Kit, repair, voltage regulator - Comprising - 1 A7809 Armature 1 A7796 Nut, adjusting 1 A7797 Screw, adjusting 2 A7802 Screw 1 A7806 Spring 3 A1666 Washer 3 51532 Washer

A.Short. [signature] for Brigadier,
(C.A. GEYTON)
Director of Mechanical Engineering.

Issue 1, 2nd March 1945.

COPY **Nº** 1669

BODIES

Instruction No. **B**

C.14-5

ALLIED LAND FORCES HEADQUARTERS

D.M.E. Technical Instruction

M.V. MODIFICATION

EQUIPMENT:—TRAILERS, 2-WHEELED, JEEP (AUST.).

SUBJECT:—Fitting of retaining pins to door catches.

REFERENCE:—L.M. 1117.

---

1. *Purpose*:—To prevent catches from unlocking.
2. *Drawings Required*:—Drg. MGO(SK)705 reproduced herein.
3. *Items Affected*:—
   - (a) New Parts:—Pins, retaining—Two (2) per Equipt.
   - (b) Replacement Parts:—Nil.
   - (c) Substitute Parts:—Nil.
   - (d) Discarded Parts:—Nil.
   - (e) Supply of Parts:—To be obtained locally.
   - (f) Material:—¼-inch M.S. rod; No. 16 Gauge Jack chain 8-inch long.
4. *Special Tools, Jigs, or Fixtures Required*:—Nil.
5. *Instructions*:—
   - (a) Drill 5/16-inch hole through door catches and angle side tops, when catch is in locked position.
   - (b) Fit 3-inch retaining pin.
   - (c) Secure retaining pin with 8-inch of jack chain.
   - (d) Spot weld end of chain to angle side top.
6. *Remarks*:—Log Books will be endorsed on completion.

FILE UNDER—"BODIES."

Rankhead

Brigadier,
Director of Mechanical Engineering.

Issue 1, 31st July, 1943.

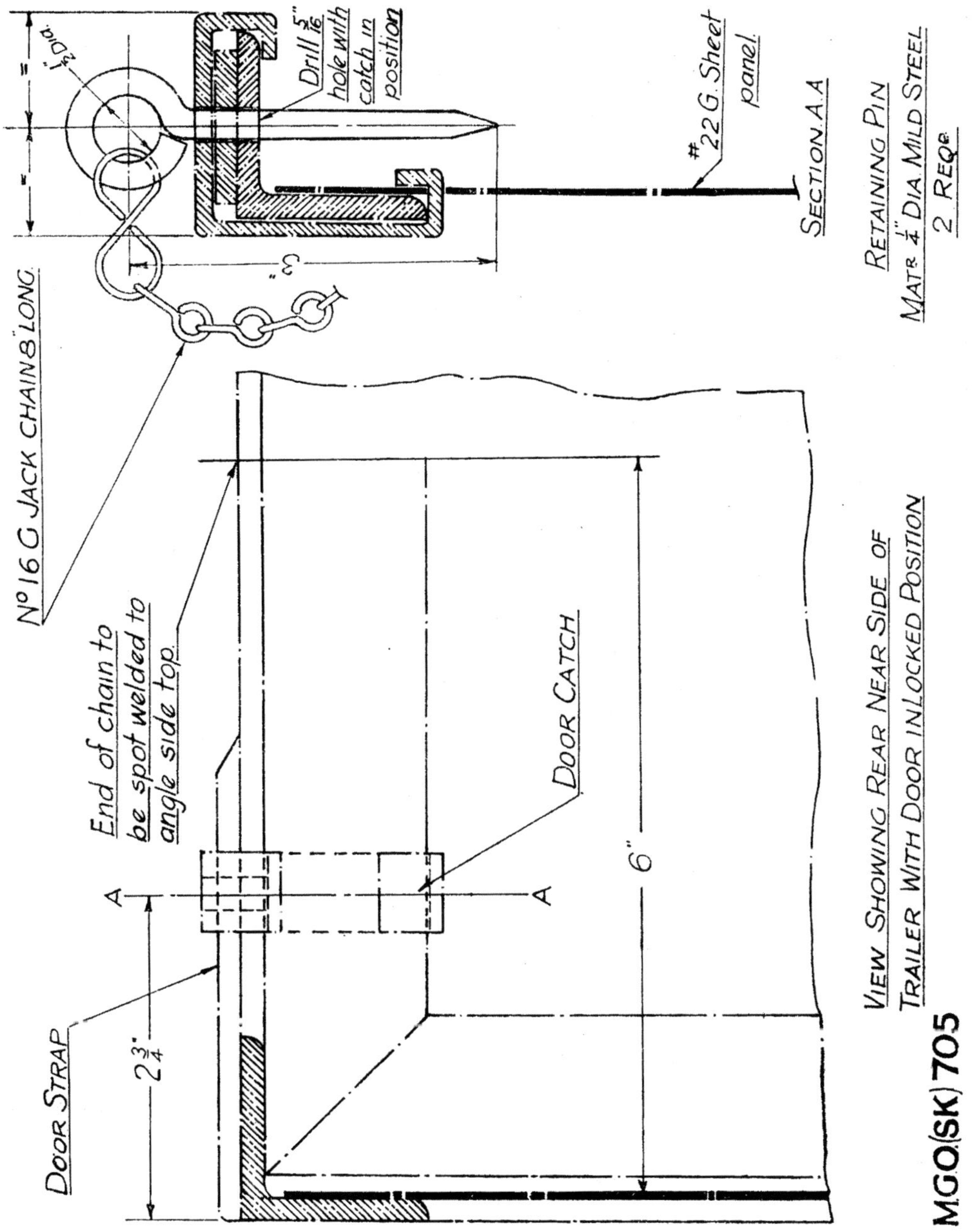

Drill 5/16" hole with catch in position
1/2" dia.
3"
#22 G. Sheet panel.
SECTION A.A
RETAINING PIN
MATR 1/4" DIA MILD STEEL
2 REQD
No 16 G JACK CHAIN 8" LONG
End of chain to be spot welded to angle side top.
DOOR CATCH
A
A
6"
DOOR STRAP
2 3/4"
VIEW SHOWING REAR NEAR SIDE OF TRAILER WITH DOOR IN LOCKED POSITION
MGO(SK) 705

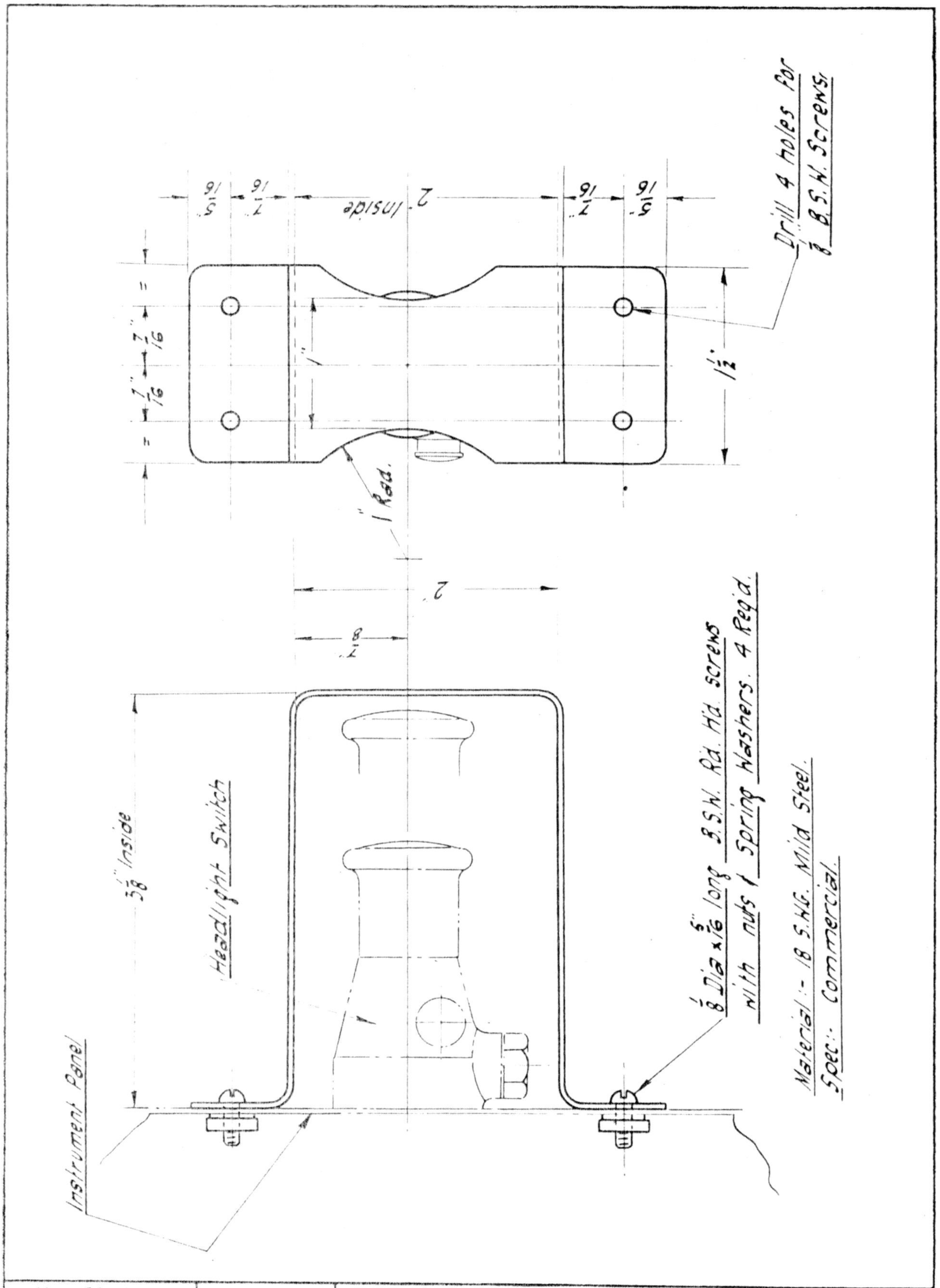

ISSUED BY	TITLE	HEADLAMP SWITCH COVER	DRAWN	REF.
DIRECTORATE OF TECHNICAL SERVICES R.A.A.F.	COMPONENT OF	JEEP	TRACED	M.R.
	MACHINE		CHECKED	R.F.
	ENGINE		APPROVED	J.R.W.
	TECH ORDER	M/T MAINTENANCE INST. Nº	DRAWING Nº	
	ISSUE Nº		C.8931	
	DATE	22.10.45		

ABOUT THE AUTHORS

Michael K Cecil

Michael K Cecil was born in Melbourne in 1955 and has been interested in military vehicles for as long as he can remember. He acquired his first vehicle – a Ford F15 CMP – in the late 1970s, which was followed by several other vehicles of various makes and models including several jeeps.

After completing university degrees in science and arts, he spent many years in both public and private consulting positions in Victoria. It was his long-held interest and research in military technology - particularly vehicles and ordnance - that led him to join the staff of the Australian War Memorial in 2000. He held various curatorial positions, becoming the Head of Military Heraldry and Technology in 2004, the position he held until his retirement in 2010. He now lives in the USA, with his wife, artist Krystii Melaine, where he continues to research and write. He has published extensively including 15 books and around 200 articles.

Keith Webb

Born in Melbourne in 1955, Keith Webb has been interested in military vehicles and aircraft for most of his life, having bought his first CMP at the age of 15. After 45 years and about 15 CMPs later, he is still keenly researching and restoring CMP vehicles and currently restoring a Canadian Military Pattern Ford FGT No9 (Aust).

Following studies in photography and film making, Keith established video production company Image Control Pty Ltd in 1982. He and his team have since produced hundreds of corporate videos, television commercials and documentary films, including extensive work for the Temora Aviation Museum and RAAF History and Heritage Branch. He has produced a lasting visual legacy of the largest military vehicle rally in Australia. Held annually at Corowa, Keith has produced a DVD of the event's highlights for almost 20 years, capturing vehicles in action and interviewing their owners.

Keith has been married to Belinda since 1981 and has four adult children and four grandchildren.